Agricultural Water Security

Agricultural Water Security: A Global Perspective offers a comprehensive exploration of the challenges and solutions surrounding sustainable water use in agriculture. Bridging the gaps between water availability, demand, and governance, this book examines the intricate connections between agricultural production, food security, and socioeconomic factors influencing water management decisions. Through global case studies and cutting-edge research, it presents strategies for balancing water supply and demand, improving water-use efficiency, and adapting to climate change and population growth. An indispensable resource for students, researchers, policymakers, and professionals in water and agricultural sciences, this book provides the insights needed to tackle the pressing issue of agricultural water security in an era of increasing water stress.

- Explores the physical, economic, and policy dimensions of agricultural water security, integrating global case studies and the latest research.
- Presents solutions for balancing water supply and demand, enhancing water-use efficiency, and adapting to climate change and population growth.
- Examines agricultural water security at farm, regional, national, and global levels, emphasizing integrated and sustainable management approaches.

Agricultural Water Security
A Global Perspective

Zied Haj-Amor

CRC Press
Taylor & Francis Group
Boca Raton London New York

CRC Press is an imprint of the
Taylor & Francis Group, an **informa** business

Designed cover image: Shutterstock

First edition published 2026
by CRC Press
2385 NW Executive Center Drive, Suite 320, Boca Raton, FL 33431

and by CRC Press
4 Park Square, Milton Park, Abingdon, Oxon, OX14 4RN

CRC Press is an imprint of Taylor & Francis Group, LLC

ISBN: 978-1-041-11550-2 (hbk)
ISBN: 978-1-041-11547-2 (pbk)
ISBN: 978-1-003-66052-1 (ebk)

DOI: 10.1201/9781003660521

Typeset in Times
by KnowledgeWorks Global Ltd.

Contents

Preface

As agriculture is currently the largest water user and its demand for water is expected to continue rising to feed a growing population, assessing current and future agricultural water security is therefore essential to anticipate potential water scenarios and develop effective adaptation strategies that enhance water productivity and help avert food and water crises. In this context, this book examines the current state and future prospects of agricultural water security by offering a comprehensive framework that explores the complex interconnections between water availability, demand, agricultural production, food security, and the socio-economic and governance factors influencing water management decisions. It analyses agricultural water security across multiple spatial scales, from individual farms to regional, national, and global levels. Drawing on hundreds of case studies from over 70 countries alongside the latest research, it provides valuable insights into the physical, economic, and policy dimensions of agricultural water security, discussing strategies for balancing water demand and supply, improving water-use efficiency, and addressing the challenges posed by population growth, climate change, and increasing water stress. Building on this comprehensive analysis, the book is an indispensable reference for students, researchers, engineers, water specialists, environmental scientists, agricultural professionals, and policymakers committed to designing effective and context-specific solutions to the complex challenges of water management in agriculture.

About the Author

Dr. Zied Haj-Amor has extensive experience in the fields of agricultural water management, soil management, climate change, and sustainable agriculture. He is the author of two books, numerous book chapters, and many research papers published in leading international journals, including Agricultural Water Management, Science of the Total Environment, and the Journal of Water and Climate Change. In recent years, he has actively contributed to several Tunisian and African research projects focused on water and soil conservation under climate change. Furthermore, he collaborates closely with researchers across Africa and Europe to enhance the productivity of saline water and soils in the face of climate change.

Acknowledgments

The author gratefully acknowledges the insightful feedback provided by his colleagues at Sfax University (Tunisia), which significantly contributed to the development of this book. Their thoughtful comments and constructive suggestions throughout the writing process were essential in enhancing the quality and depth of the work.

About the Book

Agricultural Water Security: A Global Perspective provides a comprehensive framework for understanding agricultural water security at local, regional, and global scales, addressing its physical, economic, and policy dimensions. It examines the complex interactions between water availability, demand, agricultural production, food security, and governance. Drawing on hundreds of case studies from over 70 countries and recent research, the book offers insights and strategies to balance water supply and demand in the agricultural sector, enhance water-use efficiency, and tackle challenges such as population growth, climate change, and rising water stress.

Overall, this book serves as a valuable reference for graduate students and academic staff in civil, agricultural, and water resources engineering. It is also highly useful for water resources planners and managers, flood and drought preparedness agencies, irrigation district managers, and others involved in water resources planning and management.

Acronyms and Abbreviations

AEWS	Agricultural Economic Water Scarcity
AFSs	Agroforestry Systems
AHP	Analytic Hierarchy Process
AI	Artificial Intelligence
AIWW	Actual Irrigation Water Withdrawals
ANNs	Artificial Neural Networks
ARS	Agricultural Research Service
ASR	Aquifer Storage and Recovery
AWSI	Agricultural Water Stress Index
BOD	Biological Oxygen Demand
BRB	Banas River Basin
BW	Blue Water
BWA	Blue Water Availability
BWF	Blue Water Footprint
BWS	Blue Water Scarcity
CAP	Common Agricultural Policy
CDI	Conventional Deficit Irrigation
CFSR	Climate Forecast System Reanalysis
CH$_4$	Methane
CNNs	Convolutional Neural Networks
CO$_2$	Carbon Dioxide
CR	Capillary Rise
CRM	Canadian Rocky Mountains
CT	Conservation Tillage
CWR	Crop Water Requirements
CWP	Crop Water Productivity
DA	Differential Analysis
DEM	Digital Elevation Model
DI	Deficit Irrigation
DP	Deep Percolation
E	Evaporation
E$_a$	Water Application Efficiency
ESA	European Space Agency
ET	Evapotranspiration
EU	European Union
EWS	Economic Water Scarcity
FLW	Food Loss and Waste
FAO	Food and Agriculture Organization
GCM	Global Climate Model
GCAM	Global Change Analysis Model
GBM	Gradient Boosting Machine
GHGs	Greenhouse Gases

GHM	Global Hydrological Model
GIS	Geographic Information Systems
GPS	Global Positioning Systems
GP	Great Plains
GWP	Green Water Productivity
GWR	Groundwater Recharge
GWS	Green Water Scarcity
HRUs	Hydrologic Response Units
IDF	Rainfall Intensity-Duration-Frequency
IGC	International Grains Council
IPCC	Intergovernmental Panel on Climate Change
IWD	Irrigation Water Demand
IWRM	Integrated Water Resources Management
IWUE	Irrigation Water Use Efficiency
IWMI	International Water Management Institute
IoT	Internet of Things
LDCs	Least Developed Countries
LSTM	Long Short-Term Memory
LSA	Local Sensitivity Analysis
LULC	Land Use and Land Cover
MAE	Mean Absolute Error
MADM	Multi-Attribute Decision-Making
MDB	Murray–Darling Basin
ML	Machine Learning
MLR	Multilinear Regression Model
MENA	Middle East and North Africa
N	Nitrogen
N_2O	Nitrous Oxide
NCW	Non-Conventional Water
NIR	Net Irrigation Requirements
NLP	Natural Language Processing
NSE	Nash–Sutcliffe Efficiency
O_2	Oxygen
P	Phosphorus
PDEs	Partial Differential Equations
PET	Potential Evapotranspiration
PRD	Partial Root Zone Drying
PRMS	Precipitation-Runoff Modelling System
PWP	The Permanent Wilting Point
Q	Pumping Rate
R^2	Coefficient of Determination
RDI	Regulated Deficit Irrigation
RFs	Random Forests
RMSE	Root Mean Square Error
RWH	Rainwater Harvesting

RWUE	Rainfall Water Use Efficiency
SA	Sensitivity Analysis
SC	Specific Capacity
SCE	Snow Cover Extent
SDSM	Statistical Downscaling Model
SDG	Sustainable Development Goal
SEA	Southeast Asia
SOMs	Self-Organizing Maps
SPAC	Soil-Plant-Atmosphere Continuum
SPC	Soil-Plant Continuum
SPI	Standardized Precipitation Index
SPEI	Standardized Precipitation Evapotranspiration Index
SSA	Sub-Saharan Africa
SMSs	Soil Moisture Sensors
SRTM	Shuttle Radar Topography Mission
SVMs	Support Vector Machines
SW	Surface Water
SWAT	Soil and Water Assessment Tool
SWI	Seawater Intrusion
SWMP	Stanford Watershed Model IV
SWE	Snow Water Equivalent
TP	Tibetan Plateau
TOPSIS	Technique for Order Preference by Similarity to Ideal Solution
UA	Uncertainty Analysis
UAE	United Arab Emirates
UK	United Kingdom
UPA	Urban and Peri-Urban Agriculture
USA	United States of America
USDA	United States Department of Agriculture
VW	Virtual Water
WF	Water Footprint
WFD	Water Framework Directive
WFEV	Economic Value of Water Footprint
WMO	World Meteorological Organization
WP	Water Productivity
WQI	Water Quality Indicator
WSI	Water Scarcity Indicator
WSM	Weighted Sum Model
WUE	Water Use Efficiency
ψ	Water Potential

General Introduction

The total volume of water on Earth is estimated at 1.4 billion km³ (Meran et al., 2021), yet only about 35 million km³, roughly 2.5%, is freshwater (Petersen et al., 2019). Of this small fraction, only a limited portion is accessible for human use. Despite its scarcity, freshwater is vital for sustaining both natural ecosystems and human activities. Among these, the agriculture sector, particularly agricultural production, is the largest consumer of this precious resource (Dotaniya et al., 2023). Globally, farmers depend on reliable supplies of both blue and green water to grow crops, raise livestock, and support food production systems.

However, the demand for freshwater is intensifying. Climate change, population growth (expected to reach 10.1 billion by 2050, United Nations, 2019), accelerated urbanization, increasing food consumption, the expansion of irrigated areas, and rapid economic development (El Kharraz et al., 2012; Morton, 2015) have raised pressing questions: Will rainfed and irrigated agriculture have enough water to produce sufficient food in the coming decades? What climatic and non-climatic factors must be addressed to strengthen agricultural water security?

Currently, many agricultural regions are already facing an unprecedented water crisis (La Jeunesse et al., 2016; Fiorillo et al., 2021; Teferi et al., 2025), and the situation is expected to worsen. By 2050, 87 of 180 countries, including some considered water-rich, are expected to experience water scarcity (Baggio et al., 2021). Agricultural water resources are increasingly threatened by rising demand, deteriorating water quality (Du Plessis, 2022), and climate-induced stresses. Combined with non-climatic drivers such as poor governance and unsustainable resource use, this situation could result in serious food and water insecurity, particularly in arid and semi-arid regions.

Water and food security are central to economic growth, environmental sustainability, ecosystem conservation, public health, and political stability. The inability of countries to secure these essentials could lead to severe social, environmental, and economic disruptions (Aligholi & Hayati, 2022). As a result, there is growing awareness among farmers, policymakers, researchers, and the general public that agricultural water insecurity is a rising global threat. This underscores the urgent need for sustainable freshwater management and effective adaptation strategies to prevent a global water crisis.

Before developing such strategies, however, it is essential to assess current and future trends in agricultural water supply and demand across local, regional, and global scales. A comprehensive understanding of the drivers, dynamics, and risks to agricultural water systems is essential for designing effective policies. This has led researchers to conceptualize agricultural water security, identify its key dimensions, including availability, accessibility, stability, and use, and develop indicators and advanced modelling tools to support assessment and decision-making.

Given the above discussion, this book has two primary objectives:

1. To provide up-to-date data on current and future agricultural water security at local, regional, and global levels.
2. To present actionable adaptation strategies that have been successfully implemented in diverse settings around the world, offering guidance for improving agricultural water security in the decades to come.

Drawing on hundreds of case studies from over 70 countries, this book presents a comprehensive examination of agricultural water security, from fundamental concepts to innovative solutions. It is structured in seven chapters, each designed to build the reader's understanding of the topic and support informed decision-making.

Chapter 1: Interlinkages Between Water, Agriculture, Climate Change, and Food Security: Key Facts
This chapter introduces the foundational concepts needed to understand agricultural water security, including water withdrawals, consumption, supply, demand, and distinctions between green, blue, and grey water. It lays the groundwork for the chapters that follow.

Chapter 2: Indicators of Agricultural Water Security: Advances and Limitations
This chapter reviews the key indicators used to measure agricultural water security at global, national, and local levels. It critically examines their strengths, limitations, and role in supporting effective water management.

Chapter 3: Contribution of Hydrological Modelling to the Assessment of Agricultural Water Security under Climate Change
Focusing on case studies from around the world, this chapter explores how advanced hydrological and basin-scale models can forecast future risks and support assessments of climate-related impacts on agricultural water systems.

Chapter 4: Barriers to Agricultural Water Security: A Global Analysis of Climatic Risks and Non-Climatic Constraints
This chapter investigates the major barriers to achieving agricultural water security, including both climatic challenges (e.g., droughts, floods, heatwaves) and non-climatic constraints such as governance failures, infrastructure gaps, and socio-economic inequalities.

Chapter 5: Recent Trends in Agricultural Water Security: A Global, Regional, and Local Outlook
Here, the book evaluates current trends in agricultural water use, with a focus on sustainability, irrigation efficiency, water conservation, and balancing food production needs with environmental protection.

Chapter 6: Future Trends in Agricultural Water Security under Climate Change: Global, Regional, and Local Perspectives
Based on the latest projections, this chapter examines future water availability and demand across regions, including analyses of both blue and green water resources under different climate scenarios.

Chapter 7: Adapting for Tomorrow: Strategies to Enhance Agricultural Water Security

The final chapter presents practical strategies and adaptation measures to improve water use efficiency, resilience, and equitable access to water in agriculture. Drawing on international best practices, it outlines how countries can safeguard water and food security in a changing climate.

Together, these chapters offer a holistic view of the issue, combining theoretical foundations with practical insights.

1 Interlinkages between Water, Agriculture, Climate Change, and Food Security
Key Facts

1.1 INTRODUCTION

Freshwater accounts for only 2.5% of the global water supply (Petersen et al., 2019), yet it is the most essential element for life on Earth. Thanks to the water cycle, freshwater is replenished regularly. However, because water and climate are closely interconnected, climate change over the last five decades has significantly disrupted this cycle through rising air temperatures and shifting rainfall patterns, leading to more frequent extreme weather events such as droughts and floods (Malek et al., 2018). According to the World Meteorological Organization (WMO), since 2000, the number of droughts and floods worldwide has increased by 29% and 134%, respectively, compared to the period 1980–1999 (WMO, 2020). This has resulted in global economic losses of approximately $365 billion (Lee et al., 2020).

On farms, where water, climate, and agriculture are closely interconnected, disruptions to both local and global water cycles have resulted in altered agricultural water demand, reduced water availability in many regions, increased soil erosion, and other challenges, ultimately reducing agricultural production worldwide (Haj-Amor et al., 2023). If climate mitigation efforts are too slow or insufficient, studies (e.g., Borrelli et al., 2020; Fiorillo et al., 2021) predict that these changes will continue to intensify in the coming decades at both global and local levels. This, in turn, will threaten water security, pose challenges to the sustainable management of water resources, and limit agriculture's ability to feed a global population projected to grow from 7 billion in 2010 to 9.4 billion by 2050 (United Nations, 2019), increasing global food demand by about 56% over this period (van Dijk et al., 2021). As a result, appropriate adaptation measures must be implemented urgently to avoid water and food insecurity in the coming decades. Both water and food security play a central role in supporting economic growth, environmental sustainability, public health, and political stability, meaning that the inability of some countries to ensure them could lead to severe environmental, political, social, and economic impacts (Aligholi & Hayati, 2022).

In light of the rising demand for scarce water resources and increasing food needs, a better understanding of the interlinkage between water, agriculture, climate

DOI: 10.1201/9781003660521-1

change, and food security is essential. This knowledge is crucial for assessing the likely impacts of climate change on agriculture, identifying the major challenges that may constrain water and food resources, and designing effective adaptation measures that can simultaneously reinforce water and food security under changing climatic conditions (Balasubramanya & Stifel, 2020). The objective of this chapter is to review the current information on this interlinkage. It is structured into three sections: The first describes the water cycle, what it is, how it is affected by climate change, and its role in agriculture. The second section discusses the interlinkage between water and agriculture. The final section focuses on food security, what it is, how climate change affects it, and its connection to water security.

1.2   THE WATER CYCLE

1.2.1   What Is the Water Cycle?

Water is an essential element for agriculture and life, as it impacts every aspect of our existence, from food production to public health. It covers about 70% of the Earth's surface, compared to only 30% for continents and islands (Eslamian et al., 2018). Thanks to the water cycle, water is one of the most renewable resources on Earth. As reported in Pimentel et al. (2004), the water cycle, also known as the hydrological cycle, is the continuous movement of water on, above, and below the Earth's surface.

Six major processes are involved in the water cycle: Evaporation, evapotranspiration, condensation, precipitation, surface runoff, and infiltration (Figure 1.1). These processes enable water to change from one state to another. First, heat and wind cause surface water from rivers, streams, lakes, and oceans to evaporate, forming water vapor that rises into the atmosphere. Globally, about 577×10^{12} m³ water is

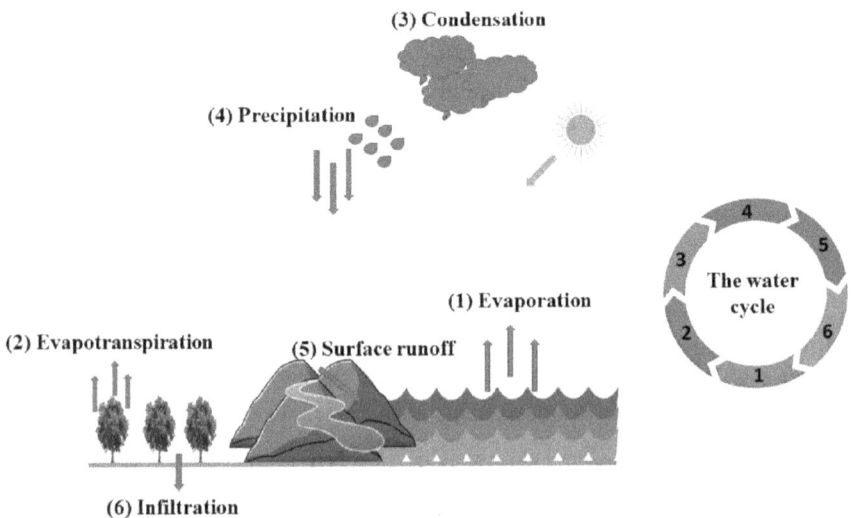

FIGURE 1.1   The major processes involved in the water cycle.

evaporated annually from the Earth's surface due to solar insolation (Pimentel et al., 2004). Additionally, water can evaporate from the soil's top layers and transpire from plant leaves, a process known as evapotranspiration. The higher the water vapor rises into the atmosphere, the more heat it loses, and the more it cools off. When the atmosphere is sufficiently cold, the water vapor freezes into tiny crystals and forms a cloud. This is called condensation.

Due to gravity, water falls back to Earth as rain, sleet, snow, or hail, a process referred to as precipitation. Depending on where it falls, water can follow different paths. If it falls on plant canopies, precipitation is intercepted by the plants, preventing it from reaching the soil. This process is known as rainfall interception. Some coniferous trees, for example, can intercept up to 35% of annual rainfall. If precipitation falls on surface water bodies (e.g., rivers or lakes), it will eventually evaporate and return to the atmosphere. However, if it falls on high mountains, it may freeze and form snow and ice. When this snow or ice melts, the water either runs off into streams, rivers, or lakes or infiltrates into the ground, replenishing vegetation and aquifers. Much of the water ultimately flows back into the ocean.

To complete the water cycle, when the groundwater level reaches the land surface, water seeps from the ground and flows into surface water bodies, allowing the entire cycle to begin again with the evaporation process (Pimentel et al., 2004).

1.2.2 The Water Cycle and Climate Change

Overall, climate change has accelerated both temporal and spatial changes in the water cycle (Yang et al., 2021). Specifically, climate change is altering the water cycle by influencing where, when, and how much water is available. Rising temperatures cause an increase in evaporation, leading to greater amounts of moisture circulating throughout the atmosphere and intensifying precipitation. During heavy rainfall events, much of the water runs into streams and rivers, resulting in more frequent and severe flooding disasters. The frequency of intense floods has increased dramatically over the last four decades. From 2001 to 2018, a total of 2900 intense flooding disasters occurred globally, and this number is expected to rise significantly in the coming decades (Lee et al., 2020). This trend makes it increasingly difficult to decide how best to protect people and infrastructure from floods.

On the other hand, increased evaporation rates also accelerate the drying of land surfaces, intensifying the severity of droughts. Surface drying reduces the amount of water that infiltrates into the ground, leading to decreased groundwater recharge (Yang et al., 2021). Globally, compared to the year 2000, the number of droughts has increased by approximately 30% and is expected to rise in the coming decades (Haj-Amor et al., 2023). As a result, communities may become more vulnerable to declining water supplies.

It should be noted that the impacts of climate change on the water cycle can significantly affect water resources management (Yang et al., 2021). For example, increased precipitation intensity may place additional stress on dams, storm drains, and flood prevention infrastructure. Furthermore, droughts can lower surface water levels, reducing water availability, disrupting water supply, and increasing abstractions from underground aquifers (Petersen-Perlman et al., 2022). Abstractions that exceed natural recharge rates can prolong groundwater deficits, reduce streamflow, and impact long-term water storage (Petersen-Perlman et al., 2022).

1.2.3 THE WATER CYCLE AND AGRICULTURE

The water cycle plays a key role in farming by providing plants and crops with a constant supply of freshwater, regulating temperatures through evaporation, providing moisture to crops, and recycling nutrients (Gettelman & Rood, 2016). At the same time, agriculture is essential to the functioning of the water cycle. Water is a primary component in farming, and plants perform numerous important functions that support the cycle. First, plants transpire water, contributing significantly to the emission of water vapor into the atmosphere via the evapotranspiration mechanism. Additionally, vegetation enhances the soil's ability to retain moisture. Plants also use their root systems to prevent water runoff and erosion. Furthermore, through photosynthesis, plants absorb carbon dioxide (CO_2), water, and sunlight to grow, releasing oxygen (O_2) into the atmosphere, which helps maintain the water cycle (Yang et al., 2021).

However, many modern farming practices can disrupt the hydrological cycle, leading to water-related concerns that threaten agricultural water security (Yang et al., 2021). For instance, in many regions, farmers have replaced natural vegetation with crop cover and pasture, significantly reducing the amount of water that infiltrates the soil and increasing runoff. Additionally, agricultural production consumes vast amounts of water, putting considerable stress on local water resources and diminishing freshwater availability needed for the proper functioning of the water cycle. Moreover, the intensive use of chemical fertilizers and the discharge of sewage can contaminate water resources, damage ecosystems, and further disrupt the water cycle (Sand-Jensen, 2013).

1.3 THE INTERLINKAGE BETWEEN WATER AND AGRICULTURE

1.3.1 ROLE OF WATER IN AGRICULTURAL PRODUCTION

As reported by Meran et al. (2021), the total volume of water on Earth is estimated at 1.4 billion km^3. Of this, only 35 million km^3, approximately 2.5%, is freshwater, and only a small fraction of that is accessible for human use (Petersen et al., 2019). Despite its limited availability, freshwater is of great importance and plays a crucial role in human activities, climate regulation, and ecosystem services (Figure 1.2). Among all human activities, the agriculture sector, particularly agricultural production, is the largest consumer of this limited freshwater resource (Dotaniya et al., 2023).

Over the past few decades, irrigation, along with improved seeds, fertilizers, and pesticides, has played a crucial role in increasing agricultural production, particularly in Asia, which is the largest user of water for agriculture globally due to its high population density (4.7 billion) and extensive rice cultivation. Rice, in particular, requires large amounts of water. Roughly 1673 L of water are needed to produce just 1 kg of rice (Mekonnen & Hoekstra, 2012).

```
┌─────────────────────────────────────────────┐
│                 Water on Earth                │
└─────────────────────────────────────────────┘
         ↓                        ↓
┌───────────────────┐   ┌───────────────────┐
│     Seawater      │   │     Freshwater     │
└───────────────────┘   └───────────────────┘
         ↓                        ↓
 The majority of water    Small fraction of the total water
                          Essential for ecosystem services
                          Essential for life
```

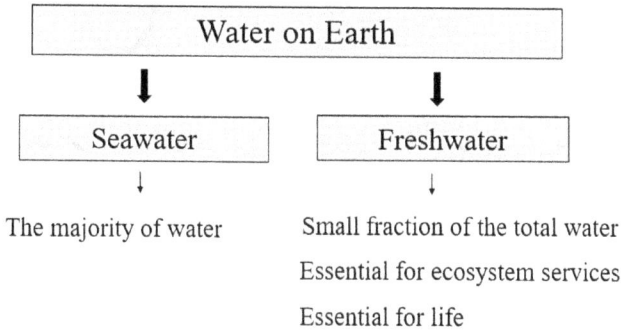

FIGURE 1.2   Earth's water distribution and the essential roles of freshwater.

In the coming decades, as climate change leads to higher air temperatures, lower soil moisture, extended periods of drought, and increased crop water demand, irrigation is expected to play a more significant role in maintaining soil moisture during dry periods, balancing crop water demand, preventing crop failures, and ultimately boosting agricultural production (Alejo & Alejandro, 2022). However, in this context, water should not be considered just an input taken for granted in agriculture. It must be managed effectively to contribute to climate change mitigation and water resource conservation. This management must consider not only water quantity but also water quality and the diverse agricultural uses of water (Balasubramanya & Stifel, 2020).

1.3.2 Importance of Water to Plants

All plants require water for photosynthesis, growth, and reproduction. The water used by plants is nonrecoverable, as some becomes part of the plant's chemical makeup, while the remainder is emitted into the atmosphere. Crops vary significantly in the amount of water (as soil moisture) required for their growth and development. For instance, rice needs a soil moisture level of at least 80%, while potatoes only require 25% soil moisture (Pimentel et al., 2004).

Plants develop a complex network of roots to continuously absorb water and transport nutrients from the surrounding soil, using them to produce food through photosynthesis. Osmosis, the natural movement of water molecules from a high-concentration area to a low-concentration area, is the mechanism by which plants absorb water from the soil (McElrone et al., 2013). Less than 5% of the water absorbed by the roots is used for plant development. The remaining 95% evaporates directly into the atmosphere through microscopic pores on the surfaces of plant leaves and stems called stomata (McElrone et al., 2013). This process is known as transpiration (Figure 1.3). The average global transfer of water into the atmosphere by vegetation transpiration from terrestrial ecosystems is estimated to account for 64% of all precipitation that falls on Earth (Pimentel et al., 2004).

However, it is important to note that this mechanism is not just a stage in the plant's life cycle. It is a vital physiological activity that directly influences plant

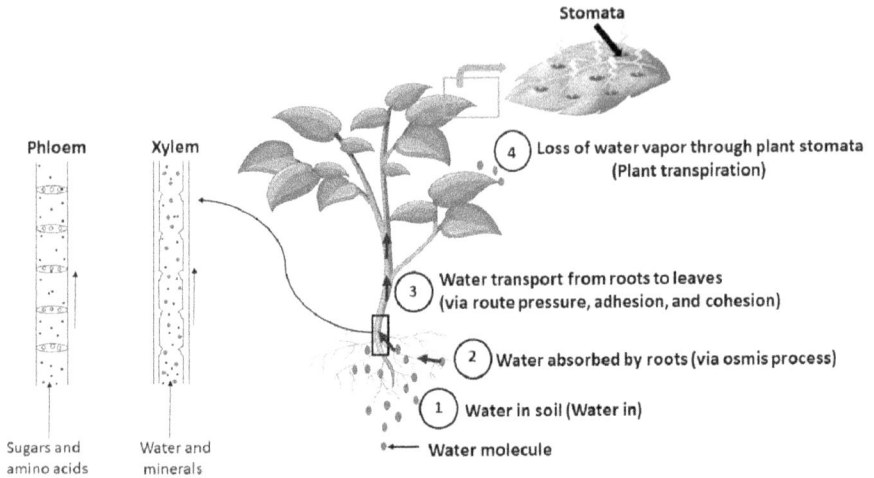

FIGURE 1.3   Water movement from plant roots to the atmosphere.

growth. Transpiration plays a central role in conducting water and minerals to the entire plant, providing a cooling effect, maintaining optimal osmotic balance, and enhancing the rigidity of leaves, roots, and other plant organs (Mengel et al., 2001). Moreover, without transpiration, excess water would accumulate in plant cells, causing the cells to burst and the plant to die (Mengel et al., 2001).

Many factors, both external and internal, can influence the rate of plant transpiration. As shown in Table 1.1, internal (or cellular) factors mainly include leaf orientation, leaf water potential (LWP), the amount of water available in a plant's leaves,

TABLE 1.1
Factors Affecting Plant Transpiration Rate

Factors	Specific Factor	Effect	References
Internal factors	Leaf orientation	Vertical leaf orientation reduces transpiration	Körner (2013)
	Leaf water potential	Reduced potential decreases transpiration	Vesala et al. (2017)
	Leaf structure	A thicker cuticle reduces transpiration	Drake et al. (2013)
	Number of stomata	More stomata increase transpiration	Drake et al. (2013)
External factors	Air temperature	Higher temperature increases transpiration	Sadok et al. (2021)
	Relative humidity	Increased humidity increases transpiration.	Mahajan et al. (2008)
	Light intensity	Lower light intensity reduces transpiration	Feng et al. (2019)
	Carbon dioxide (CO_2)	Lower CO_2 levels increase transpiration	Agrawal and Deepak (2000)
	Wind speed	Higher wind speed reduces transpiration	Schymanski and Or (2016)
	Water supply	Limited water supply reduces transpiration	Koehler et al. (2023)

leaf structure, and the number of stomata on a leaf. External (or environmental) factors, on the other hand, include air temperature, relative humidity, light intensity, atmospheric CO_2 concentration, wind speed, and water supply (Tuzet, 2011). These factors interact in a complex way, making it essential to fully understand how plants and crops respond to them. This knowledge is crucial for predicting plant growth and yield under various environmental conditions and for providing guidance on ensuring sufficient agricultural production (Zhu et al., 2022).

Many mathematical equations can be used to describe transpiration rate changes as a function of different environmental factors (Zhu et al., 2022). The Penman-Monteith equation (Eq. 1.1) is one of the most commonly used ones because it requires only conventional observations as input (Monteith, 1965):

$$LE_t = \frac{\Delta(R-G)+\rho\, C_p \dfrac{d}{r_a}}{\Delta+\gamma\left(1+\dfrac{r_s}{r_a}\right)} \tag{1.1}$$

To resolve the equation mentioned above, the input variables as shown in Table 1.2 are required.

In addition to their central role in plant transpiration, stomata also play a key role in photosynthesis, the process by which plants convert atmospheric CO_2 into more complex organic compounds, particularly glucose, using the sun's energy (Tuzet, 2011). The role of stomata in photosynthesis can be summarized as follows: First, stomata allow plants to take in CO_2 from the atmosphere, which is essential for photosynthesis. Then, using the sun's energy, plants combine the absorbed CO_2 with water to produce glucose and release O_2, a waste product of the photosynthesis process (Tuzet, 2011). This chemical reaction typically takes place in the chloroplasts found in the inner layers of plant leaves (Tuzet, 2011).

TABLE 1.2

Input Data Required to Calculate Transpiration Rate (E_t)

Symbol	Symbol Name	Unit
L	Latent heat of vaporization of water	J kg^{-1}
ρ	Dry air density	kg m^{-3}
c_p	Specific heat of the air at constant pressure	J kg^{-1} K^{-1}
d	Air saturation deficit	Pa
r_a	Bulk aerodynamic resistance	s m^{-1}
r_s	Canopy resistance	s m^{-1}
γ	Psychrometric constant	kg m^{-2} s^2 K^{-1}
Δ	Temperature derivative of the saturated vapor pressure curve	K^{-1}

Note: E_t (transpiration rate) is expressed in kg m^{-2} s^{-1}

1.3.3   Soil-Water-Plant Relationship

The soil-water-plant relationship defines the status of water in the soil-plant-atmosphere continuum (SPAC). Quantifying this relationship is crucial for determining the water status in a plant relative to its needs, understanding how much water the soil can hold, formulating optimal irrigation schedules, and developing better soil management practices for promoting optimal plant growth. Practically, the soil-plant-water relationship can be described using three key indicators: Water potential, water movement, and water content, each of which highlights a particular aspect of this relationship. These indicators are explained in greater detail in the following subsections.

1.3.3.1   Water Potential

Water potential (ψ) refers to the potential energy of water per unit volume relative to pure water (under atmospheric pressure and ambient temperature). It is typically expressed in units of pressure, such as megapascals (MPa). Plants rely on water potential to move water from the soil to the leaves and other plant organs, facilitating processes such as transpiration, nutrient uptake, photosynthesis, and cell expansion. It is a useful concept for understanding the movement and function of water within plants and their surrounding environment (soil and atmosphere). Water potential describes both the direction (downward gradient) and the flow rate of water transport within the soil-plant continuum (SPC) (Steppe, 2018).

Water potential reflects the "strength" with which the plant holds water and quantifies the tendency of water to move from the soil to the plant due to factors like gravity, osmosis, pressure, and other mechanical forces known as matrix effects (Vos & Haverkort, 2007). Therefore, for a given SPC, the total water potential (Ψ_{system}) can be estimated using the following equation (Eq. 1.2):

$$\Psi_{system} = \Psi_g + \Psi_s + \Psi_p + \Psi_m \tag{1.2}$$

where Ψ_g is the gravity potential, Ψ_s is the solute potential (or osmotic potential), Ψ_p is the pressure potential, and Ψ_m is the matric potential. The following are some details about these potentials:

- *Gravity potential* (Ψ_g): This refers to how quickly water flows in the plant and how much energy the plant must expend to move water against gravity. Gravity always pulls water down into the soil, which reduces the total potential energy of the water in the plant (Ψ_{system}). The taller the plant, the higher the water column, and consequently, the greater the effect of Ψ_g (Giménez et al., 2013).
- *Osmotic potential* (Ψ_s): This reflects the reduction in the free energy content of water molecules due to the presence of dissolved solutes in the cell sap. A higher concentration of dissolved solutes in the cell sap results in fewer water molecules being free to move, thus making Ψ_s more negative (Giménez et al., 2013). Ψ_s is often measured using thermocouple psychrometry or hygrometry.

- *Pressure potential* (Ψ_p): This refers to the hydrostatic pressure exerted on water within a cell. It is typically measured using an osmometer.
- *Matric potential* (Ψ_m): This represents the reduction in water potential due to the interaction of water with the soil's solid matrix, particularly through capillarity and adsorption mechanisms. It is always negative because it reduces the mobility of water molecules.

Water moves from an area of a higher Ψ_{system} to an area of a lower Ψ_{system}.

1.3.3.2 Water Movement in Soils

When water is supplied to soil, either through rain or irrigation, it first infiltrates through the large open pores between soil particles, called macropores, and then fills the smaller pores, known as micropores. When both macro- and micropores are fully saturated with water, the soil is considered "water-saturated." At this saturation stage, water begins to drain downward freely due to gravity and a lack of superficial tension. This portion of water that drains out freely, not retained by the soil, is termed gravitational water. Usually, a small amount of this water is available to plants because it drains quickly from the soil macropores toward the water table.

Once all the gravitational water has drained, the soil reaches field capacity, and plant roots begin to search for water that is not influenced by gravity or superficial tension, known as capillary water. Capillary water is water held in the soil micropores with a tension of approximately −33 kPa during periods when no water is being supplied (e.g., no rain or irrigation). This is the primary source of water for plants; without it, plants cannot survive or grow. The term "field capacity" is crucial because it represents the maximum amount of water the soil can hold. In temperate climates, capillary water rises to the surface of the soil, while in arid climates, it evaporates before reaching the surface, leaving the upper soil layers dry.

As time progresses, both soil evaporation and plant transpiration further reduce the water content in the soil until water no longer moves in response to capillary forces. At this point, all the capillary water has been extracted by plant roots, and the remaining water is bound tightly to the soil particles by immense physical pressure, typically ranging from −1500 to −10,000 kPa. This bound water is known as hygroscopic water. The Permanent Wilting Point (PWP) occurs when soil particles hold the water so tightly that plant roots can no longer absorb it. At this stage, the plant has extracted all available water. Unlike capillary water, hygroscopic water has high viscosity and elasticity and can only be removed as water vapor by drying the soil at 105°C. If plants are exposed to this "hygroscopic" condition for extended periods, they begin to exhibit reduced growth and yield. In such cases, additional water supply to the soil becomes urgently necessary to restore plant health.

The aforementioned forms of water in the soil are graphically shown in Figure 1.4.

1.3.3.3 Water Movement In Plants

Water movement from the roots to other parts of the plant is essential for its survival. Like other multicellular organisms, plants rely on a vascular system to transport water and nutrients. Without it, plant cells would be deprived of essential resources.

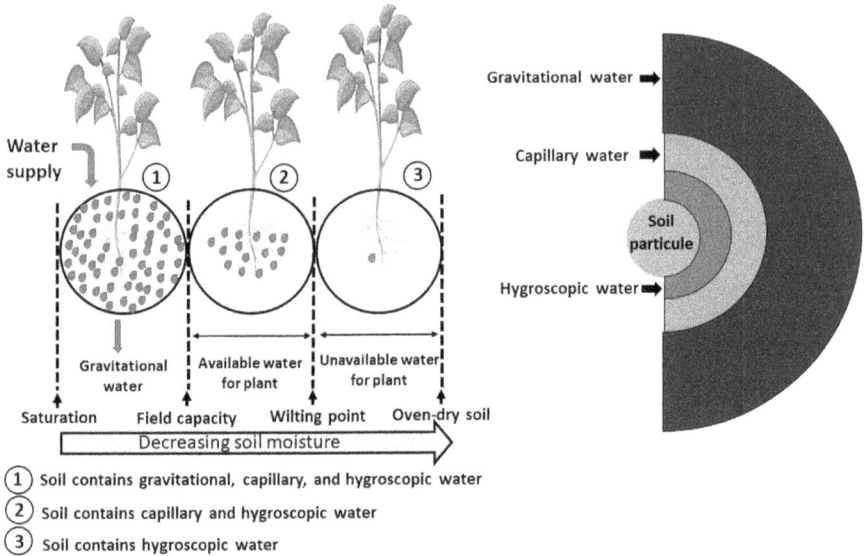

1. Soil contains gravitational, capillary, and hygroscopic water
2. Soil contains capillary and hygroscopic water
3. Soil contains hygroscopic water

FIGURE 1.4   The main forms of water that can be found in soil.

The vascular system consists of specialized tissues that serve the dual purpose of supplying cells with water and nutrients while removing waste. In plants, the vascular system is primarily composed of two types of tissue: Xylem and phloem.

The xylem transports water and mineral ions from the roots to other parts of the plant, based on the cohesion-tension theory. Meanwhile, the phloem is responsible for transporting sugars and amino acids through the source-to-sink theory (McElrone et al., 2013).

The xylem consists of dead, water-conducting cells known as tracheary elements, which are thickened with spiral structures. The phloem, on the other hand, consists of living sieve tube cells that are connected by sieve plates to form continuous tubes (Figure 1.3). These cells play a critical role in filtering water at the root level and have a higher resistance to water flow compared to the xylem (McElrone et al., 2013).

Water first enters the plant through the roots via a process known as osmosis. Osmosis is the movement of water across a semipermeable membrane, from areas of high-water concentration to areas of low water concentration. In the SPC, the soil typically has a higher concentration of water than the plant roots, driving the water into the roots.

Once inside the roots, water is transported through the xylem to other parts of the plant through three major mechanisms: Root pressure, adhesion and cohesion, and transpiration (Figure 1.3).

- *Route pressure*: Route pressure is a result of the movement of water molecules from the soil into the xylem. This process causes the xylem to have a higher concentration of dissolved minerals than the rest of the plant. As a consequence, water moves passively from the plant roots, which have the highest concentration, to the stem and leaves, which have the lowest concentration. This movement of water, along with dissolved minerals, is called root pressure, and it helps to push the xylem sap (a solution of water and minerals) upwards into the stem and leaves.

- *Adhesion and cohesion*: Water molecules in the xylem exhibit two important properties: Adhesion and cohesion. Adhesion refers to the tendency of water to stick to the walls of the xylem vessels, while cohesion refers to the strong attraction between water molecules due to hydrogen bonds. These cohesive forces can reach up to 30 MPa, enabling water molecules to form a continuous column. This continuous column of water helps move water upward from the roots to the rest of the plant.

- *Transpiration*: Transpiration is the process by which excess water in the plant is lost to the atmosphere as water vapor. This primarily occurs through the stomata of the leaves. Transpiration not only helps in water regulation but also plays a vital role in maintaining the flow of water and nutrients from the roots to the upper parts of the plant.

1.3.3.4   Soil-water Content

Soil-water content, commonly known as soil moisture, expresses the amount of water contained in the soil. It can be expressed on either a mass or volume basis. On a mass basis, it is referred to as gravimetric water content (Eq. 1.3), while on a volume basis, it is known as volumetric water content (Eq. 1.4):

$$\theta_g = \frac{M_w}{M_{ds}} \tag{1.3}$$

where θ_g is the gravimetric water content at a particular soil depth (g^3 g^{-3}), M_w is the mass of water (g), and M_{ds} is the mass of dry soil (g).

$$\theta_v = \frac{V_w}{V_s} \tag{1.4}$$

where θ_v is the volumetric water content at a particular soil depth (cm^3 cm^{-3}), V_w is the volume of water in the soil (cm^3), and V_s is the total volume of soil (cm^3).

On farms, volumetric water content should be regularly measured to understand the amount of water in the soil relative to plant requirements. This information helps farmers make informed decisions regarding irrigation, allowing them to accurately control water application based on their crops' needs.

Various laboratory and field methods are available to measure soil moisture content, but the most commonly used method is soil moisture sensors (SMSs). This method is effective, efficient, and provides a rapid, direct measurement of soil moisture in the field.

Many SMSs function as variable resistors, which are resistors whose resistance level changes based on an environmental factor. In the case of SMSs, their resistance levels vary with the amount of moisture they come into contact with. Typically, an SMS has two long conductors, called electrodes, separated by a certain distance.

These electrodes allow electricity to flow freely. Since water is a good conductor of electricity, when these electrodes are inserted into the soil (which contains water), the conductivity of the soil increases, and the resistance level of the sensor decreases. Thus, the amount of water in the soil can be indirectly determined by measuring the soil's electrical conductivity and applying Ohm's law (Eq. 1.5):

$$V = R \times I \tag{1.5}$$

where V is the voltage measured across the conductor (V), R is the resistance of the conductor (Ω), and I is the current through the conductor (A).

1.3.4   AGRICULTURAL WATER USES

Globally, agriculture consumes about 85% of total water use annually (D'Odorico et al., 2020), compared to only 15% for industrial activities and urban use (FAO, 2019). However, it is important to note that the share of agricultural water use differs significantly between countries and regions. For instance, the European Union (EU) has the smallest share of agricultural water use, around 30%, whereas South Asia, one of the world's most densely populated regions, has the largest share, approximately 90% (FAO, 2019).

In the context of agriculture, it is crucial to distinguish between "water use" and "water consumption." Water use refers to the total amount of water used by farmers for specific purposes, such as irrigation or livestock farming, within a given area. On the other hand, water consumption is the portion of water use that has been evaporated, transpired, or incorporated into products or crops (Rost et al., 2008). Overall, agricultural water is primarily used for irrigation, livestock production, and aquaculture.

1.3.4.1   Irrigation

Irrigation is the largest consumer of freshwater, accounting for 70% of global withdrawals. However, it also suffers from significant water losses, with 60% of the total withdrawal wasted due to inefficient irrigation systems, poor application techniques, and the cultivation of water-intensive crops that exceed the water availability in their growing environment (Biswas et al., 2025).

Over the last three centuries, the total irrigated area worldwide has expanded from only 5 million hectares (Mha) in 1700 to more than 338 Mha in 2018 (Martínez-Valderrama et al., 2023). The largest increase in irrigated areas has been primarily observed in North Africa, South Asia, Central Asia, and the Middle East, where rainfall is very low, soil moisture deficit is high, and staple food crops require large and regular amounts of water to produce agricultural yields (Wu et al., 2017; Martínez-Valderrama et al., 2023). For example, from 1980 to 2016, the area under irrigation in India (South Asia) increased by 72% to maintain the same level of agricultural production each year (Dinar et al., 2019). However, due to climate change, irrigation is expected to become a more essential practice in other regions such as Central Europe and Eastern Asia (Mancosu et al., 2015).

Over the past five decades, irrigation has played a crucial role in stabilizing crop yields and intensifying agricultural production, leading to significant improvements in food security and nutrition in many countries (FAO, 2019). Globally, irrigated agriculture accounts for only 19% of total farmland but contributes more than 40% of global food production, providing many socioeconomic benefits (Hanjra & Qureshi, 2010). This important contribution of irrigated agriculture to the world's food supply is expected to continue in the coming decades, driven by further warming, ongoing changes in precipitation patterns, and increasing food demand. Fortunately, there is significant potential to expand irrigated areas in regions where sufficient water is available, especially in South America and Sub-Saharan Africa (SSA) (Mancosu et al., 2015). However, it is critical to manage irrigation water efficiently to minimize losses and ensure its sustainable use.

1.3.4.2   Livestock Production

In addition to its central role in irrigated agriculture, water is essential in livestock watering, feedlots, and dairy operations. Livestock, such as cattle, poultry, goats, sheep, horses, and pigs, require drinking water for optimal health and to produce healthy products such as milk, eggs, and meat. Water is also necessary to clean and sanitize dairy equipment, cool livestock housing, and manage animal waste effectively (Dieter et al., 2018).

Globally, about 41% (4387 km³) of total agricultural water is used for livestock production (Heinke et al., 2020; FAO, 2019). A significant portion of this water is used to produce feed crops, forages, and grazed grass for animals. Cattle farming is the largest water user in this sector, accounting for 32% of total livestock consumptive water use, followed by dairy cows (18%), industrial pigs (14%), layer poultry farming (4%), smallholder chicken farming (3%), and smallholder pigs (2%) (Heinke et al., 2020).

1.3.4.3   Aquaculture

Aquaculture, which involves raising aquatic organisms (e.g., fish, shellfish) or growing plants in freshwater or brackish water systems under controlled conditions for food consumption, is a rapidly growing food sector, especially in South and Southeast Asia (SEA). It provides employment and food security for millions of people in these regions (Tezzo et al., 2021). World aquaculture production has increased from less than 43 million metric tons (Mt) in 2000 to more than 120 Mt in 2019 (Verdegem et al., 2023). Asia, particularly South and Southeast Asia, is the largest aquaculture producer, contributing about 90% of global production, followed by the Americas (3%), Europe (2%), and Africa (2%) (Verdegem et al., 2023).

Aquaculture is a water-intensive activity, as water is the primary medium in which fish and plants are produced (Troell et al., 2014). Water use in aquaculture can be divided into consumptive water use (which reduces stream flow and removes water in the form of biomass) and total water use, which includes rainfall, runoff, infiltration, and management inputs (Troell et al., 2014). Globally, the overall use of water by aquaculture systems is estimated to range from 70 to 390 km³ per year (Troell et al., 2014). Although this volume is relatively small compared to irrigated agriculture and livestock production, it is crucial to minimize water use in aquaculture

```
┌─────────────────────────────────────────────────────┐
│                    Water types                        │
└─────────────────────────────────────────────────────┘
        ↓                  ↓                  ↓
┌───────────────┐  ┌───────────────┐  ┌───────────────┐
│  Green water  │  │   Blue water  │  │   Grey water  │
└───────────────┘  └───────────────┘  └───────────────┘
        ↓                  ↓                  ↓
 Soil moisture available  Surface water   Treated wastewater
    from rainfall         Groundwater     suitable for reuse
```

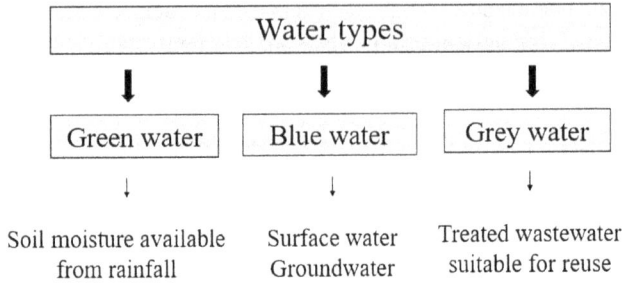

FIGURE 1.5   Types of water used in agriculture: Green, blue, and grey.

to reduce pressure on freshwater resources (Verdegem et al., 2023). Specifically, promoting water recycling in land-based systems, reducing water consumption, and facilitating nutrient recovery and reuse should be prioritized (Verdegem et al., 2023).

1.3.5   SOURCES OF WATER FOR AGRICULTURE

Globally, about 8362 km³ of water are used by agriculture every year. Of this total volume, approximately 80% is green water, 11% is blue water, and 9% is grey water (Hoekstra & Mekonnen, 2012). All these types of water can be used in agriculture (Figure 1.5), and each type requires specific management.

1.3.5.1   Green Water

Green water refers to the portion of precipitation that is first stored in the soil as soil moisture and then used by plants and crops via evapotranspiration for growth and development (Johansson et al., 2016). Therefore, crop green water consumption can be defined as the evapotranspiration from soil moisture replenished by precipitation in cropland (Huang et al., 2019a). On average, approximately 60% of precipitation ends up as green water (Ahmed et al., 2022).

Green water scarcity in rainfed croplands occurs when the total amount of rainfall is insufficient to meet crop water requirements (CWR), potentially leading to critical water stress during crop growth (He & Rosa, 2023). To mitigate this stress, supplemental blue water is often provided through irrigation (He & Rosa, 2023). In recent decades, increasing temperatures have led to an overall decrease in soil moisture from rainfall (Ahmed et al., 2022). A continued decline in soil moisture could heighten the need for supplemental irrigation in rainfed agriculture, resulting in smaller yields and potentially significant impacts on food production, as the productivity of rainfed crops largely depends on green water (He & Rosa, 2023).

1.3.5.2   Blue Water

Agricultural blue water refers to water withdrawn from groundwater and open-surface water bodies (rivers, lakes, wetlands, and reservoirs) for specific agricultural uses, such as irrigation or livestock farming (Rost et al., 2008). There are many large blue-water resources on Earth, with the largest being groundwater, lakes, and rivers.

1.3.5.2.1   Groundwater

Groundwater is the most abundant and easily accessible source of freshwater, followed by lakes and rivers. Research shows that over 30% of all freshwater resources on Earth are stored as groundwater. The total volume of groundwater is more than 100 times the volume available in lakes and rivers. Globally, 980 km³ of water are extracted annually from groundwater, 70% of which is used for agriculture (Margat & van der Gun, 2013). The five countries with the largest annual groundwater extractions are India, China, the USA, Pakistan, and Iran (Margat & van der Gun, 2013).

1.3.5.2.2   Lakes

Lakes are one of the most important surface water resources, accounting for about 0.3% of the world's readily available water supply. Many freshwater lakes are located in high-altitude regions, with nearly 50% of the world's lakes found in Canada alone. Many lakes, especially in arid regions, become saline through evaporation, concentrating the inflowing salts. Major salt lakes include the Caspian Sea, the Dead Sea, and the Great Salt Lake.

1.3.5.2.3   Rivers

A river basin is an area of land where water drains to a common outlet, such as a river or lake (Molden et al., 2011). The total volume of water in the world's rivers is estimated at 2115 km³ (Groombridge and Jenkins, 1998).

In-depth monitoring of the annual flows of the top ten major river basins (i.e., Mississippi, Nile, Congo, Amazon, Amur, Yangtze, Paraná, Yenisey, Ob-Irtysh, and Yellow rivers) from 1948 to 2004 revealed that more rivers showed decreasing trends due to climate change (Lakshmi et al., 2018). The most significant changes in the flow regime were observed during the spring, coinciding with snowmelt (Nijssen et al., 2001). The sharpest declines in river flows were noted in the Middle East, North Africa, South Asia, and the western USA (Lakshmi et al., 2018). For instance, during this period, the flow of the Nile River in Egypt (North Africa) has been reduced by about 115 m³/s per decade (Mahmoud et al., 2022).

A river basin is considered an open system when it provides enough water to satisfy withdrawal needs while maintaining its ecological services. It is considered a closing system when water use reaches or exceeds the discharge volume (Molle et al., 2007). Over the last few decades, heat waves, frequent droughts, and a lack of precipitation have worsened this issue, depriving millions of river-dependent people of full access to water. A prominent example is the Colorado River in North America, where a combination of overuse and a historic drought in the early 21st century has reduced the river's flow by 20% compared to the 20th century. This puts about 40 million people who rely on this river at high risk of water insecurity. According to Milly and Dunne (2020), the annual flow of the Colorado River has dropped by about 9% for every 1°C of warming. The study suggests that, without cuts to greenhouse gas emissions, the river's discharge could shrink by up to 31% by 2050. This has driven diverse groups (e.g., American Rivers and International Rivers) to advocate for the protection of river basins, improve water access, and help maintain waterway flow.

The expansion of irrigated areas is leading to a substantial decline in blue water availability worldwide (Ahmed et al., 2022). Blue water scarcity in irrigated croplands arises when blue water resources cannot meet irrigation water needs (He & Rosa, 2023). Currently, about 50% of irrigated croplands face critical blue water scarcity, which is expected to affect about 59% of the world's population by 2050 (Rockström et al., 2009; He & Rosa, 2023). Given that global agricultural production needs to increase by 56% by 2050 to feed 9.4 billion people, the blue water shortage must be addressed through effective solutions (Ahmed et al., 2022). Changing global cropping patterns, rainwater harvesting, desalination, water reuse, and water conservation could be helpful solutions to address the blue water shortage (Chouchane et al., 2020).

1.3.5.3   Grey Water

Grey water refers to wastewater produced from various household activities, such as showering, bathing, laundry, and dishwashing, excluding contributions from toilet flushing (Al-Jayyousi, 2003). In recent decades, increasing pressure on conventional water resources from rapidly growing agricultural water demands and climate change has led many farmers worldwide to irrigate crops, particularly non-food crops, with treated grey water. Irrigating with treated grey water can offer several benefits, including conserving fresh water resources, particularly in water-scarce areas, providing nutrients like nitrogen (N) and phosphorus (P) to crops, thus potentially reducing the need for synthetic fertilizers in agriculture, and supporting the food security of families (Helmecke et al., 2020).

Despite these potential advantages and the fact that the suitability of grey water for irrigation has been rigorously addressed to ensure crop growth, health, and environmental safety, concerns remain regarding its use in irrigated agriculture. Many people are still uncertain about the possibility of soil and crop contamination, as well as the potential effects on human health (Silva et al., 2023). Therefore, the local use of grey water in irrigated agriculture requires a shift in public perception regarding the reuse of wastewater (Silva et al., 2023).

1.4   THE INTERLINKAGES BETWEEN WATER, AGRICULTURE, FOOD SECURITY, AND CLIMATE CHANGE

1.4.1   WHAT IS FOOD SECURITY?

Despite the fact that sufficient food is produced to supply all 8 billion people globally, about 815 million individuals, roughly one in every nine, do not consume enough food to maintain good health (Dinar et al., 2019). These individuals are not confined to less developed nations; malnutrition is a global issue.

Food security refers to the ability to provide enough food for people. Based on the definition proposed by the United Nations (UN), food security exists when all people, at all times, have physical, social, and economic access to sufficient, safe, and nutritious food for an active and healthy life. For each country, food security is crucial for ensuring a healthy and well-nourished population that can actively perform different daily activities.

As reported by de Oliveira Veras et al. (2021), food security is a multidimensional concept that encompasses four key dimensions: Food availability (whether there is sufficient healthy and acceptable food), accessibility (whether individuals can acquire the available food through reliable resources), utilization (whether individuals can properly use the available food to achieve optimal nutritional status), and stability (which refers to the consistency of the food supply over time). Food insecurity occurs when one or more of these dimensions are compromised.

There are many countries around the world where food security is a significant challenge. Several key factors contribute to this issue.

1.4.1.1 Population Growth

The global population has rapidly increased in recent decades, from 6 billion in 1999 to 8 billion in 2022, driving higher food demands. At the same time, available land and resources (such as water) for farming are becoming scarcer. This combination of factors makes it increasingly difficult to produce enough food to meet everyone's needs.

1.4.1.2 Changes in People's Diets

As countries develop, dietary patterns are shifting, with people consuming more food, particularly meat. China is a prime example, where population growth and rising household incomes have led to a significant increase in meat consumption, especially pork, poultry, beef, and mutton, making it the largest consumer of meat globally for many years (Liu et al., 2017a). However, meat production uses more water than crop cultivation, and the growing demand for meat exacerbates the pressure on the world's freshwater resources. This is particularly concerning, as 29% of the agricultural sector's total water footprint is tied to the production of animal products, especially beef (Mekonnen and Hoekstra, 2012). Reducing meat consumption could help decrease freshwater use and support food security.

1.4.1.3 Climate Change

Climate change is affecting agricultural areas, with some experiencing increased droughts while others face more frequent flooding (Haj-Amor et al., 2023). Both droughts and floods are linked to soil degradation processes such as erosion, salinization, and acidification, which weaken soil health, impair crop growth, and reduce the capacity of agricultural land to provide food (Blanc & Noy, 2023). As a result, yields of food crops such as wheat, rice, maize, barley, and soybeans are decreasing. For example, global wheat yields are decreasing by an average of 1% per year due to climate change (Ray et al., 2019). Additionally, farmers are now facing new pests and diseases that were not previously a concern, posing risks to crops and livestock (Bett et al., 2019).

Regarding the four dimensions of food security, without adequate intervention, climate change is anticipated to lead to food insecurity in the following ways:

- *Food availability*: Reduced crop yields and soil degradation.
- *Food stability*: Increased climate variability affects food production consistency.
- *Food access*: Higher food prices and lower incomes, particularly in agriculture-dependent regions like SSA and South Asia.
- *Food utilization*: The spread of plant and animal diseases due to climate change (Mirzabaev et al., 2023).

1.4.1.4   Conflict

In conflict-affected areas, food production often declines significantly, and infrastructure that provides water may be destroyed. The transportation of food is frequently disrupted. A relevant example is the Russia-Ukraine conflict. According to the International Grains Council (IGC), Ukraine's grain production in the 2022–2023 agricultural season decreased by about 29% compared to the previous season, with further reductions anticipated. As one of the world's top grain exporters, producing 12% of global wheat exports, this conflict has put around 1.7 billion people at risk of hunger and 276 million individuals in severe food insecurity (Lin et al., 2023). The countries most affected include those heavily dependent on Ukrainian wheat, such as Egypt, Mongolia, Georgia, Azerbaijan, and Turkey (Lin et al., 2023).

1.4.2   How Does Climate Change Affect Food Security?

Climate change has the potential to significantly impact food security, both directly and indirectly. These are some examples of how climate change can affect food security.

1.4.2.1   Changes in Crop Yields

Climate change can lead to changes in crop yields due to shifts in temperature and rainfall patterns, as well as an increased frequency and intensity of extreme climate events such as droughts and floods. These changes can reduce food production, decrease food availability and accessibility, and drive-up food prices. The number of climate change-related disasters, such as extreme heat, droughts, and floods, has doubled since 1990, leading to substantial reductions in crop yields (Haj-Amor et al., 2023). On a global scale, Zhao et al. (2017) found that a 1°C increase in global average temperature could reduce global yields of wheat, rice, maize, and soybean by approximately 6%, 3%, 7%, and 3%, respectively. On a regional scale, Nguyen et al. (2020) revealed that drought accounts for about 30% of year-to-year yield changes in maize and soybean in the southeastern USA, with the risk of crop yield loss increasing from 50% to 80% as drought intensity shifts from moderately dry to extremely dry.

1.4.2.2   Changes in Food Quality

Climate change can also affect the quality of food, as temperature and rainfall changes influence the composition and nutritional value of crops. This can result in lower food quality, reducing the nutritional value of diets and increasing the risk of micronutrient deficiencies. Specifically, exposure to higher levels of atmospheric CO_2 can diminish the nutritional content of major food crops like rice, wheat, and corn. Increased CO_2 levels can dilute the overall mineral content of plant tissues, reduce canopy transpiration, and hinder nutrient uptake by plant roots, resulting in significant decreases in iron, zinc, and other micronutrients in plants (Semba et al., 2022). Globally, around two billion people are affected by nutrition issues related to zinc and iron deficiencies (Niyigaba et al., 2019). These deficiencies are major contributors to the mortality of children and pregnant women, particularly in developing countries (Semba et al., 2022). Studies (Broberg et al., 2017; Myers et al., 2014) have shown that increased CO_2 can reduce zinc content by 9% and iron content by 7% in wheat grains (Semba et al., 2022).

1.4.2.3 Access to Food

Climate change can also affect access to food, as changes in temperature and rainfall impact individuals' ability to grow and harvest crops, as well as the transport and distribution of food. This can lead to job losses, food shortages, reduced food access, decreased purchasing power for low-income populations, and increased food prices. People most at risk are those dependent on agriculture for their livelihoods, particularly in Asia and SSA. It is estimated that up to 820 million people worldwide rely entirely or partially on agriculture for their livelihoods (HLPE, 2014). Climate-related disasters such as floods and droughts, job losses, and unsafe working conditions may force many of these individuals to migrate to urban areas, further hindering agricultural productivity and threatening food security (HLPE, 2014).

1.4.2.4 Changes in Livestock Production

Changes in temperature and rainfall can affect feed availability and quality, while also increasing the risk of diseases and parasites. This can lead to decreased animal production, reduced availability of meat and dairy products, and heightened food insecurity (Hashmi et al., 2021). For example, the warmer climate in northern Europe has facilitated the rapid spread of bluetongue disease, severely impacting sheep and cattle and causing significant socioeconomic concerns. Other transboundary diseases, such as Rift Valley Fever and Lumpy Skin Disease, are becoming increasingly difficult to control and predict due to climate change (Bett et al., 2019).

1.4.3 How Does Climate Change Affect Agriculture?

Overall, climate change can have two distinct types of effects on agriculture. First, one of the primary causes of climate change, CO_2 emissions, may benefit plants by enhancing photosynthesis. This phenomenon is known as the CO_2 fertilization effect. Second, altered temperature and rainfall patterns can significantly impact crop and livestock productivity. These two types of effects are discussed in the following subsections.

1.4.3.1 CO_2 Fertilization Effect

Due to the combustion of fossil fuels, the concentration of CO_2 in the atmosphere has increased from 280 ppm in 1800 to approximately 410 ppm in 2020, and it is projected to reach 750 to 1300 ppm by 2100 (Allan et al., 2021). For many crops, particularly C3 crops (e.g., wheat, rice, and soybeans), this elevated CO_2 concentration is not considered a "pollution issue." Instead, it acts as a form of fertilizer, as it leads to an increased rate of photosynthesis. In other words, alongside the water molecules (H_2O) absorbed from the surrounding soil, the elevated CO_2 concentration allows plants to obtain enough CO_2 molecules from the atmosphere to form glucose ($C_6H_{12}O_6$) through the photosynthesis process in the leaves (Eq. 6):

$$6\ CO_2 + 6\ H_2O + \text{Sunlight} \rightarrow C_6H_{12}O_6 + 6\ O_2 \qquad (1.6)$$

This leads to increased plant growth as glucose is used by plants as a main source for metabolic energy. Many studies all over the world have correlated increasing CO_2 with increasing crop yields. One of the best sources for this correlation can be found at "CO_2 Science" (http://www.co2science.org/). Overall, increasing CO_2 from 475 to 600 ppm can increase leaf photosynthetic rates and crop yields by about 40% and 17%, respectively (Taub, 2010).

Increased CO_2 concentrations also allow plants to photosynthesize at a lower stomatal conductance, which reduces transpiration rates and prevents unnecessary water loss into the atmosphere. This enhances leaf-level water-use efficiency and increases plant resistance to drought, particularly in water-limited regions. Previous studies have shown that elevated CO_2 levels can help mitigate or even prevent the effects of water stress in various crop species, including soybean (Qiao et al., 2010), wheat (Allen et al., 2011), tomato (Liu et al., 2019), and pepper (Fan et al., 2020). Elevated CO_2 can benefit plants by promoting osmotic adjustment, providing additional carbon for root expansion, and enabling roots to absorb more water from deeper soil layers during periods of water deficit (Fan et al., 2020).

However, it is important to note that several factors may limit the benefits of the CO_2 fertilization effect. Some of these include:

- *Nutritional quality decline*: Long-term exposure to elevated CO_2 levels may lead to an increase in plant carbohydrate accumulation at the expense of proteins and minerals (e.g., iron, zinc, and magnesium), thereby reducing the nutritional quality of food crops. This phenomenon, known as "CO_2 acclimation," may reduce protein and mineral concentrations in plants by 5% and 15%, respectively (e.g., Myers et al., 2014; Loladze, 2014). Such declines in key nutrients could elevate the risk of malnutrition. It is projected that these reductions in zinc and iron concentrations alone could result in around 125 million disability-adjusted life-years globally, with SEA and SSA being the most affected regions (Weyant et al., 2018).
- *Increased plant diseases*: Higher CO_2 concentrations may encourage the development of plant diseases. Elevated CO_2 levels can lead to denser canopies and increased humidity, which creates favorable conditions for various plant pathogens, including foliar diseases, powdery mildews, and downy mildews (Gullino et al., 2018). Combined with the decline in nutritional quality, plant diseases pose a significant challenge to global food security in the coming decades.
- *Nutrient uptake and CO_2 release*: According to Terrer et al. (2021), the increased growth driven by the CO_2 fertilization effect requires plants to extract more nutrients from the soil. This promotes microbial activity, which, in turn, leads to the release of CO_2 into the atmosphere that would otherwise be retained in the soil.

1.4.3.2 Effects on Crop and Livestock Productivity

Over the past few decades, climate change has led to rising air temperatures, shifting rainfall patterns, more frequent and severe extreme climate events (such as droughts and floods), and rising sea levels. These changes have created increasingly difficult conditions for growing crops, resulting in higher rates of crop failure across various regions of the world. Climate change has negatively impacted the yields of several key food crops, including rice, wheat, soybeans, sugarcane, and barley. For instance, global rice yields are currently declining by about 0.3% annually, while global wheat yields are decreasing by approximately 0.9% annually (Ray et al., 2019).

However, it is important to note that not all effects of climate change are detrimental. In higher-latitude countries such as Canada and Russia, crop yields have significantly increased due to the combined effects of higher temperatures, a longer growing season, and CO_2 fertilization. On the other hand, lower-latitude countries like Kenya, Indonesia, and Brazil have experienced severe declines in crop yields as a result of reduced precipitation (Haj-Amor et al., 2023). Some projections suggest that, without adequate adaptation measures and with greenhouse gas emissions continuing at their current levels, global crop yields could decline by up to 30% by 2050 (Global Commission on Adaptation, 2019).

Climate change has also negatively impacted livestock productivity through changes in feed availability and quality, water availability, animal growth, production, disease prevalence, and biodiversity. These effects primarily result from rising temperatures, increased CO_2 concentrations, and shifts in precipitation patterns, often in combination (Rojas-Downing et al., 2017). While estimates of livestock productivity under global climate change scenarios in the coming decades are limited, some projections are available. For instance, Boone et al. (2018) anticipated a global decrease in livestock productivity of up to 9% by 2050 under a high greenhouse gas emission scenario (RCP8.5).

1.4.4 How Does Agriculture Affect Climate Change?

The relationship between agriculture and climate change is twofold. Climate change can affect agricultural production, but agriculture itself can also emit greenhouse gases (GHGs) that contribute to climate change. Agriculture is a significant source of GHG emissions, particularly methane (CH_4) and nitrous oxide (N_2O), both of which are highly effective at trapping heat in the atmosphere, contributing to further global warming. According to the Intergovernmental Panel on Climate Change (IPCC), agriculture is responsible for up to 24% of global GHG emissions, with the primary sources of these emissions are livestock sector, land use change and deforestation, rice cultivation, and energy use (Figure 1.6).

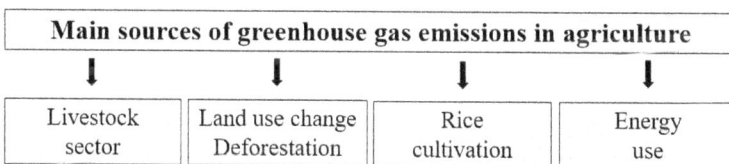

Main sources of greenhouse gas emissions in agriculture			
↓	↓	↓	↓
Livestock sector	Land use change Deforestation	Rice cultivation	Energy use

FIGURE 1.6 Main sources of greenhouse gas emissions in agriculture.

1.4.4.1   Source 1: Livestock Sector

Globally, livestock production accounts for approximately 70% of all agricultural CH_4 emissions (Bateki et al., 2023). Ruminants, such as cows, sheep, and goats, emit CH_4 during digestion through a process known as enteric fermentation. When these animals eat grass, microorganisms in their stomachs break down the fiber, converting it into CH_4, which is then emitted when they burp or pass gas. Each cow can emit about 3 tons of CO_2 equivalents of CH_4 per year (Grossi et al., 2018), making livestock digestion responsible for one-third of all anthropogenic CH_4 emissions (De Bhowmick & Hayes, 2023). In addition to ruminants, the cultivation of crops such as corn and soybeans to feed livestock also contributes to emissions, primarily from the use of inorganic fertilizers and the conversion of carbon-storing grasslands and forests into croplands. Furthermore, the decomposition of livestock manure in anaerobic conditions releases significant amounts of CH_4 and N_2O.

1.4.4.2   Source 2: Land Use Change and Deforestation

Land use change generally refers to the transformation of natural landscapes, such as rainforests and grasslands, to support human activities, particularly agriculture. As these landscapes are converted to farmland, the carbon previously stored in these ecosystems is released into the atmosphere as CO_2, N_2O, and other GHGs. Land use change accounts for approximately 42% of all agricultural GHG emissions (Poore & Nemecek, 2018), with deforestation being the major contributor. When forests are cleared for agricultural expansion, the CO_2 they have stored is released, and the land's ability to sequester carbon is lost. This creates a double burden, increasing atmospheric CO_2 levels and reducing carbon sequestration capacity. In regions such as South America and SEA, where deforestation is widespread due to rising food demand, international efforts have sought to implement reforms for better deforestation monitoring and control.

1.4.4.3   Source 3: Rice Cultivation

Rice, a staple food cultivated on 11% of global cropland (Seck et al., 2012), is one of the largest emitters of CH_4. According to the IPCC, rice cultivation contributes approximately 9% of global CH_4 emissions. In SEA, one of the world's largest rice-producing regions, rice is responsible for up to 33% of the region's CH_4 emissions. Rice is typically grown in flooded fields, which create anaerobic conditions that limit the exchange of air between the soil and the atmosphere. This environment is ideal for bacteria to decompose organic matter and release CH_4. Given that rice is both a major CH_4 emitter and a global staple, improving rice cultivation practices is essential to reducing its impact.

1.4.4.4   Source 4: Energy Use

Energy use on farms is crucial for agricultural production, but it also leads to CO_2 and non-CO_2 emissions. Farms require energy for machinery, water pumping, greenhouse and livestock stable heating, agrochemical production, and more. The generation of energy from fossil fuels (coal, oil, or gas) results in the release of significant amounts of GHGs, including CO_2 and N_2O, which trap heat in the atmosphere.

1.4.5 Water and Food Security

Water is an essential element throughout the entire food supply chain, from primary production to consumption. It plays a critical role in agricultural food production; without it, cultivating crops and raising livestock would be impossible (Miller et al., 2021). Both food crops (such as corn, wheat, and rice) and livestock (such as cows, pigs, and sheep) require substantial amounts of water to grow. On average, producing 1 kg of soybean, wheat, and rice requires about 2145, 1827, and 16,773 L of water, respectively. Furthermore, producing 1 kg of beef requires more than 15,000 L of water (Okutan & Akkoyunlu, 2021). As a result of these significant water demands, agriculture is the world's largest consumer of water.

The management of agricultural products from "farm" to "fork" significantly contributes to wastewater generation. Approximately 30% of global food production (about 1.3 billion tons) is wasted each year across the entire food supply chain, meaning the water used to produce this food is also lost (Bharathi et al., 2023). This highlights the critical need for action to reduce food loss and waste (FLW) at every stage of the food supply chain (Table 1.3) to enhance food security and optimize water use (Uhlenbrook et al., 2022; Mirzabaev et al., 2023). Substantial reductions in FLW are essential to meet the challenge of feeding a growing population and to achieve Sustainable Development Goal 2 (SDG #2), "zero hunger," which aims to end hunger by 2030.

TABLE 1.3

Main Ways to Decrease Food Loss and Waste across the Entire Food Supply Chain

Contributor	Main Action(s)
Policymakers	• Promoting educational programs to value food • Providing financial support over the entire food supply chain • Ensuring coordination between food supply chain stakeholders
Individuals	• Smart shopping • Proper food storage • Reuse leftover food
Retailers	• Promoting short food supply chains • Avoiding food expiration • Using statistical data to address food waste
Farmers	• Preventing unsuitable agronomic practices • Promoting harvesting robot systems • Promoting climate-resistant crops
Food distributors	• Avoiding rejected shipments • Ensuring safe and efficient food distribution • Promoting innovations in food distribution to support sustainability
Manufacturers	• Promoting optimal, safe, and sustainable food production systems • Enhancing energy efficiency • Producing good with less water

Agriculture contributes to climate change by emitting GHGs, particularly N_2O and CH_4, which elevate air temperatures and disrupt the Earth's water cycle (Mirzabaev et al., 2023). Rising temperatures can increase water demand, and in areas where precipitation decreases, many croplands, especially rainfed farms, will require more irrigation to grow crops, produce food, and sustain the livelihoods of farming communities. However, as water availability for irrigation is projected to decline in many regions due to climate change, it is expected that a growing number of people will face critical food insecurity in the coming decades (Misra, 2014).

Global food insecurity has been steadily increasing, primarily due to climate change and its impact on the availability of freshwater (Dinar et al., 2019). A recent study on global food security revealed that the number of chronically malnourished people rose from 777 million in 2015 to 815 million in 2016 (Dinar et al., 2019). The most severe food insecurity has been predominantly observed in SSA and SEA, driven by recurring extreme climatic events such as droughts and floods, as well as heightened water scarcity (Dinar et al., 2019). The Russia-Ukraine conflict and the ongoing effects of the COVID-19 pandemic have further exacerbated the situation (Wudil et al., 2022). These regions are also projected to be the most vulnerable to climate change-induced food shortages by 2050 (Dinar et al., 2019). Without significant investments in adaptation, it is forecasted that continued warming and reduced rainfall in SSA and SEA will cut crop yields by up to 20% and 30%, respectively, by 2050 (compared to 1990 levels). This, in turn, will intensify food insecurity (Connolly-Boutin & Smit, 2016; Aryal et al., 2020). To secure food availability by 2050, both regions must develop a clear strategy for enhancing agricultural production in the face of a changing climate. This strategy must account for all potential threats to agricultural output. Moreover, translating this vision into reality will require substantial financial investment, with the potential for significant increases in agricultural productivity and improved food security.

To feed a global population expected to grow from 8 billion in 2023 to around 9.4 billion by 2050 (United Nations, 2019), food production must rise by approximately 56% over this period (van Dijk et al., 2021). To achieve this, profound changes in policy and management throughout the entire food supply chain are urgently needed to optimize the use of available water resources and meet the increasing demand for food and other agricultural products (Uhlenbrook et al., 2022). Specifically, since water is essential not only for agricultural production but also for the entire food supply chain, the following eight measures are crucial (Uhlenbrook et al., 2022):

1. Developing smarter and more sustainable ways to produce more food with less water.
2. Implementing early warning systems for extreme climate events to protect food production from climate change.
3. Reducing GHG emissions from livestock farming.
4. Promoting agricultural development in least developed countries (LDCs) by providing the necessary skills, knowledge, and financial resources to protect water resources and ensure food security.
5. Reducing water losses throughout the agricultural production chain.

6. Advancing efficient water and food recycling strategies.
7. Encouraging a global shift to healthy diets produced by sustainable food systems.
8. Changing water use behavior among key stakeholders.

Without these essential changes, achieving food security for the anticipated 9.4 billion people by 2050 will not be possible (Uhlenbrook et al., 2022).

1.4.6 AGRICULTURAL WATER USES UNDER CLIMATE CHANGE

A deeper understanding of agricultural water scarcity is crucial for developing effective water resource management practices that are adaptable to a changing climate. Previous studies on agricultural water security have consistently shown that climate change over the next few decades will cause critical changes in the availability and consumption of both green and blue water globally, potentially threatening water and food security worldwide (Liu et al., 2022a). Changes in rainfall patterns and temperatures will directly impact the terrestrial water budget, and combined with more erratic and uncertain water supplies, climate change will exacerbate water stress in currently water-scarce regions and induce water scarcity in areas with previously abundant water resources. Specifically, Wada and Bierkens (2014) demonstrated that increasing temperatures and rising evaporation demand could lead to an 8% and 14% increase in global green and blue water consumption, respectively, between 2000 and 2050. Liu et al. (2022b) further suggested that global croplands could lose about 39% of their water by 2050. This implies that by 2050, the world could face unprecedented agricultural water insecurity. The highest agricultural water deficits are expected in North Africa, the southern edge of the Sahel, South Africa, Northeast China, and Central America, while East Asia and the Pacific are projected to experience the lowest deficits, indicating that local solutions are essential to combat water scarcity (Liu et al., 2022b).

In addition to the impacts on water quantity, climate change is also expected to worsen water quality over the next few decades, affecting both surface water and groundwater resources due to increased flooding, prolonged droughts, and accelerated sea-level rise (Treidel et al., 2011):

- Flooding can severely impact water quality by transporting a range of contaminants, including chemicals (such as fertilizers), heavy metals, and pathogens, into water bodies.
- Droughts typically cause significant drops in water levels in rivers, lakes, ponds, wetlands, and aquifers, leading to water flow from polluted surface waters into groundwater bodies. This can degrade the quality of the retained groundwater. A notable example is the Yongding River catchment in North China, where the decline in groundwater levels since 1999, exacerbated by recurring droughts, has resulted in the drying up of main river courses and groundwater pollution with industrial and domestic effluents, severely impacting agricultural activities in the Beijing Plain (Treidel et al., 2011; Jiang et al., 2014).

- Sea-level rise can lead to seawater intrusion in coastal aquifers, especially during storm surges, causing seawater to infiltrate land and contaminate groundwater. This process occurs in nearly all coastal regions and has significant consequences for soil salinity and crop production. In southern Bangladesh, for instance, many hectares of farmland are lost each year due to salinization from seawater intrusion, making coastal agriculture increasingly vulnerable (Salehin et al., 2018). The growing frequency of floods and accelerated sea-level rise in the coming years are expected to exacerbate this issue (Salehin et al., 2018).

1.5 CONCLUSIONS

This chapter provided a comprehensive understanding of the interconnectedness between water, agriculture, climate change, and food security. It highlighted that global climate change was a central factor driving numerous challenges, including extreme climate events (such as floods and droughts), disruptions in hydrologic systems, deteriorating water quality, water shortages, declining agricultural productivity, and growing food insecurity. In addition to climate change, other contributing factors, such as population growth, inefficient water use, land use changes, rising incomes, shifting dietary preferences, food loss and waste, and political conflicts, were identified as key drivers of water scarcity and food insecurity. As these challenges intensified across various regions and scales, there was an urgent need to develop science-based methodologies to identify optimal strategies for managing and protecting our finite land, soil, and water resources. Moreover, effective adaptation measures need to be urgently implemented to sustainably address the diverse water and food-related issues at hand. To strengthen water and food security, it was crucial to mitigate the impacts of climate change on agriculture, meet water demands for farming, address water scarcity in vulnerable regions, increase food production with fewer resources, and reduce food loss along supply chains.

2 Indicators of Agricultural Water Security
Advances and Limitations

2.1 INTRODUCTION

Agricultural water security is closely linked to food security, ecological health, and global economic stability. It plays a vital role in ensuring food production, addressing climate change challenges (e.g., water shortages), fostering socioeconomic development, and supporting the Sustainable Development Goals (SDGs), particularly SDG 2 (ending hunger), SDG 6 (sustainable water use), and SDG 11 (sustainable cities and communities) (Meran et al., 2021).

Assessing agricultural water security is essential for gaining a comprehensive overview of Earth's freshwater resources, identifying causes of water stress in agriculture, anticipating potential risks to agricultural production and food security, and developing sustainable water management practices (Jeyrani et al., 2021). This need has led researchers to conceptualize agricultural water security and develop several indicators to measure its key dimensions, water availability, accessibility, stability, and use. The most commonly used indicators include the Water Scarcity Indicator (WSI), Water Footprint (WF), Crop Water Productivity (CWP), Irrigation Efficiency Indicator, Water Quality Indicator (WQI), Agricultural Drought Indicator, and Resilience Indicator (e.g., Liu et al., 2022a).

In recent years, these indicators have been refined to address the complex challenges affecting agricultural water security, such as climate change, climate variability, water governance, water quality degradation, and water conflicts. For instance, Veettil and Mishra (2020) suggested estimating average water security at the county level by combining green water scarcity (GWS) and blue water scarcity (BWS) indicators. Based on the WF concept, Rosa et al. (2020) introduced an economic agricultural water scarcity indicator for croplands, which accounts for irrigation constraints due to limited institutional and economic capacity rather than hydrological limits. Additionally, Liu et al. (2022b) developed an integrated agricultural water scarcity indicator incorporating both blue and green water components to estimate global water shortages under different future climate scenarios.

Using these enhanced indicators, various studies worldwide have assessed agricultural water security (Meuwissen et al., 2019). Their findings have been adapted by decision-makers to address regional priorities and challenges (Meuwissen et al., 2019). However, comprehensive studies covering all relevant indicators for assessing agricultural water security remain limited and require further updates. To bridge this knowledge gap, this chapter reviews different indicators of agricultural water

DOI: 10.1201/9781003660521-2

security and their role in promoting sustainable water resource management at the farm, basin, national, regional, and global levels. Data from multiple levels are essential for designing effective agricultural water conservation strategies tailored to the needs and challenges of rural communities.

2.2   WHAT IS AGRICULTURAL WATER SECURITY?

Overall, water security refers to having enough safe water to meet all needs. Each individual requires approximately 1200 m³ of water annually to be water secure (Allan, 2010). Water availability (whether water is available in the physical environment), accessibility (whether water can be obtained at an affordable cost), use (whether there is sufficient clean and safe water for all needs), and stability over time are the four dimensions that make up the concept of water security (Young et al., 2021). Water insecurity arises from the compromise of one or more of these dimensions, which can be caused by issues related to water shortages, climate change, natural disasters (e.g., earthquakes), or pollution. Globally, about 1.2 billion people are currently struggling with water insecurity, with the majority of them living in arid and semiarid countries (Ingrao et al., 2023). Over the coming few decades, this number will tend to increase sharply because of human population growth and the increasing frequency and severity of extreme climate events such as droughts and floods (Ingrao et al., 2023).

The aforementioned definition of water security can be rewritten to define agricultural water security as follows: Over time, water of adequate quality should be available at a stable rate, easily accessible, and reasonably priced in order to ensure sustainable crop and livestock production (Malekian et al., 2017) (Figure 2.1).

Theoretically, Malekian et al. (2017) suggest that agricultural water security should involve the protection of vulnerable farmers from social, political, and environmental challenges that hinder their access to sufficient water for sustainable farming or restrict their ability to cultivate their land freely. Furthermore, the methods used to provide agricultural water must not harm others or degrade the environment (Harrington, 2013; Malekian et al., 2017).

Agricultural water insecurity occurs when the demand for agricultural water exceeds the total volume of accessible freshwater resources that are naturally available in a given domain (e.g., a farm, a watershed, country, or region), either as blue or green water. This implies that agricultural water insecurity can result not only from shortages of blue water (surface and groundwater) but also from shortages of

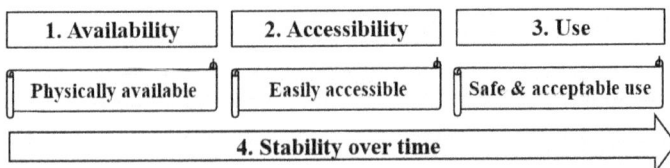

FIGURE 2.1   Dimensions of agricultural water security (availability, accessibility, use, and stability).

infiltrated rain in the soil (green water), which may restrict crop yields (Malekian et al., 2017). Overall, agricultural water insecurity is a seasonal issue because of seasonal fluctuations in water availability and quality (Marcantonio, 2018). Rosa et al. (2020) reported that about 86% of global croplands face seasonal water insecurity for at least one month per year. Given that agriculture is the largest water user globally, agricultural water security is of great importance, and the inability of countries to sustain it may have critical effects on agricultural production, food security, agricultural economic security, rural development, and much more.

On the other hand, it is worth pointing out that agricultural water security could be assessed not only through the physical availability of freshwater resources in relation to demand but also through other economic and social factors such as effective agricultural water management and planning strategies, the ability of institutions to offer water-related services, and long-term economic plans (Gain et al., 2016). As a result, agricultural water security should be quantitatively assessed taking into account its physical, socioeconomic, and governance dimensions.

In other words, agricultural water security in a given area is assessed through three key concepts: Water stress, water shortage, and water scarcity (Veettil & Mishra, 2020):

1. *Agricultural water stress*: This occurs when water demand exceeds the available supply during a specific period. It reflects the pressure placed on water resources due to farming activities, climate variability, and inefficient water use.
2. *Agricultural water shortage*: This refers to a physical lack of sufficient freshwater to meet the basic needs of farmers. It can result from prolonged droughts, over-extraction of groundwater, or declining river flows. Unlike water stress, which considers demand, water shortage focuses on the absolute availability of water.
3. *Agricultural water scarcity*: This includes both physical water shortages and economic limitations that prevent access to agricultural water. It accounts for factors such as poor water management, infrastructure deficits, and affordability. Even in regions with abundant water, scarcity can occur if water resources are not equitably distributed or properly managed.

Together, these concepts help assess the agricultural water security of a region by considering not just how much freshwater is available, but also how accessible and sustainable it is for the population.

2.3 PHYSICAL AVAILABILITY AND USE OF FRESHWATER

2.3.1 BLUE WATER

Blue water refers to freshwater found in aquifers and open surface waterbodies (e.g., rivers, lakes, wetlands, and reservoirs) that can be extracted for various agricultural uses. Approximately 30% of the world's freshwater resources are stored as blue water (Falkenmark & Rockström, 2006).

Blue water security can be defined as the ability of available blue water resources within a given domain to meet water needs, particularly for irrigation, while also providing essential ecosystem services such as climate regulation, water cycling, and soil formation (Falkenmark, 2013). Blue water insecurity arises when the demand for blue water exceeds its availability (Ingrao et al., 2023). In addition to the imbalance between water demand and availability, competition for water among users, overexploitation of surface water and groundwater, and insufficient instream flows serve as alarming indicators of critical blue water insecurity (FAO, 2012).

Given the increasing scarcity of freshwater resources driven by climate change, rapid population growth (82 million per year), agricultural intensification, and urbanization, it is essential to monitor and optimize blue water consumption across agricultural activities, particularly in irrigation (Meneguzzo & Zabini, 2021). This underscores the importance of effectively monitoring blue water consumption in agro-food systems to ensure sustainability.

2.3.1.1 Blue Water Availability

In the absence of sufficient rainfall, crops are irrigated with water drawn from blue water resources. This accounts for why agriculture consumes roughly 70% of blue water. At the same time, since blue water is "recharged" by precipitation, the spatio-temporal distribution of rainfall is the most critical factor influencing its availability (Molden et al., 2011). This highlights that decreasing precipitation due to climate change could result in a significant reduction in blue water availability (BWA). Without sufficient access to blue water resources, both water and food security cannot be achieved.

BWA is defined as the quantity of freshwater present in the following water bodies (Table 2.1).

2.3.1.1.1 Surface Water Bodies

They refer to any body of water above the Earth's surface, such as rivers, streams, lakes, reservoirs, and wetlands. These water resources are particularly abundant in regions with high rainfall and/or significant snowmelt. Below are some details regarding these resources.

TABLE 2.1

Main Resources of Blue Water

Resource		Example	Key Information
Surface water	Rivers & streams	Nile River, Egypt	It satisfies 95% of Egypt's irrigation demands
	Lakes & reservoirs	Lake Victoria, East Africa	It is a great future food basket for Africa
	Wetlands	The Everglades, South Florida	It is a unique and intricate hydrological system
Groundwater		(-)	It makes 99% of the world's liquid freshwater
Glaciers & ice caps		(-)	They store about 70% of the Earth's freshwater
Snowpack		(-)	It has a leading role in cooling climate

2.3.1.1.1.1 Rivers and Streams   Rivers and streams play a crucial role in promoting both agricultural productivity and water security due to their significant contribution to various aspects of agriculture. They are vital for irrigation (providing water for crop irrigation in areas with insufficient rainfall), hydropower production (generating electricity to power irrigation pumps and other agricultural machinery), floodplain agriculture (enabling crop cultivation in areas alongside rivers that are flooded during high-water events), transportation of fertile sediments (redistributing nutrients and organic matter beneficial to agricultural lands along riverbanks), and water storage (e.g., dams built on rivers can store water during periods of high flow and release it gradually during dry periods) (Sordo-Ward et al., 2019).

A prime example of how rivers support both water security and agricultural productivity is the Nile River in Egypt. Water from the Nile satisfies approximately 95% of Egypt's irrigation needs (Mostafa et al., 2021). Additionally, through mechanisms such as flooding, infiltration, and seepage from irrigation canals, the river contributes significantly to replenishing the Nile Delta aquifer, one of the largest aquifers in Egypt (Negm, 2018). Moreover, the Nile is essential for meeting the food needs of millions of Egyptians, as it enables the cultivation of staple crops such as wheat, rice, and maize.

2.3.1.1.1.2 Lakes and Reservoirs   Both lakes (natural systems) and reservoirs (water storages) play vital roles in controlling flood risks (by storing excess water during heavy rainfall events), protecting crops under drought stress (by providing a reliable water supply during dry periods), managing water quality (by trapping sediments and filtering out pollutants), and ensuring food security (through fish production).

Lake Victoria in East Africa, the largest lake on the continent, provides a prime example of these benefits. The lake serves as a major water storage source for three African countries, Tanzania, Kenya, and Uganda, and is one of the largest freshwater fisheries in the world, supporting the livelihoods and food security of millions. Moreover, it plays a crucial role in the irrigation of staple crops (such as rice, maize, and cassava), cash crops (like coffee, tea, and cotton), and horticultural crops (including bananas, pineapples, and mangoes), offering nutritional diversity to the people of East Africa and beyond.

Despite these positive contributions, the water of Lake Victoria must be managed with great care to avoid the risks of water scarcity and reduced crop production, particularly in the context of climate change (Hengsdijk et al., 2014). Effective management strategies for the lake's water are urgently needed, as Lake Victoria is considered a crucial future food basket, with the potential to provide agricultural products and food to both East Africa and other parts of the world (Mutharika, 2010).

2.3.1.1.1.3 Wetlands   Natural wetlands, such as marshes, swamps, and bogs, can effectively remove pollutants and excess nutrients, thereby improving water quality. This improved water quality, in turn, enhances the productivity of agricultural wetlands by promoting nutrient availability for crops, reducing soil degradation processes like salinity and erosion, and boosting microbial activity in the soil. Wetlands also have a remarkable ability to store water for surrounding agricultural areas through various natural ecological processes, including surface water retention, nutrient cycling, groundwater recharge, and sedimentation. These processes collectively contribute to agricultural water security and sustainable land management practices.

Furthermore, wetlands help mitigate downstream flooding during wet seasons and reduce the impact of severe rain events by holding water within a watershed. Despite these valuable functions, it is crucial to emphasize the need for careful management of wetland agriculture to prevent the degradation of the ecological character of wetlands (Verhoeven & Setter, 2010).

A notable example of the role wetlands play in water purification and water storage for agriculture is the Everglades in South Florida, USA. This globally recognized wetland system provides vital water to the Everglades Agricultural Areas during dry seasons. The Everglades, often referred to as a "river of grass," is a unique and intricate hydrological system characterized by slow-moving water flowing southward through a shallow, wide riverbed. This slow flow creates ideal conditions for aquatic plant growth and supports diverse wildlife. However, the Everglades' capacity to purify water and meet the water needs of surrounding agricultural areas has been significantly diminished. This is primarily due to factors such as altered rainfall patterns, increased temperatures and evapotranspiration, and high-nutrient pollution from agricultural runoff. As a result, the Everglades has been reduced to just 50% of its original size, further compromising its ability to perform these essential functions (Mitsch et al., 2015; Schade-Poole & Möller, 2016).

2.3.1.1.2 Groundwater

Groundwater is natural freshwater stored beneath the Earth's surface, within the pores of soils and rocks. The 2022 annual UN World Water Development Report showed that groundwater accounts for about 99% of the world's liquid freshwater and contributes to 25% of all water consumption for human activities. Groundwater supplies roughly 50% of drinking water, 40% of irrigation water, and 33% of industrial water (Lall et al., 2020).

In response to a changing climate, large aquifer systems, such as Australia's Great Artesian Basin and the aquifers of the Midwest in the USA, have served as vital buffers against severe drought effects. For example, in California, during dry years, groundwater meets up to 46% of the state's total water needs (Ke & Blum, 2021).

Various researchers and organizations have attempted to estimate the total volume of groundwater on Earth to understand its distribution across continents (Table 2.2). These efforts have significantly advanced our knowledge of global groundwater availability.

Based on preliminary estimates, the global volume of groundwater is roughly 23.4×10^{15} m³ (Figure 2.2). However, reliable, accurate, and up-to-date data on the volume of groundwater stored in Earth's aquifers remains elusive. This data gap has hindered the development of clear public policies that could help communities respond effectively to climate change (Dudley Ward, 2024). Therefore, it is critical to fill this gap to reduce uncertainties surrounding the future of groundwater resources in a changing climate. Access to such data would be essential for developing a comprehensive global groundwater database and enhancing the ability of communities to effectively address climate change impacts.

TABLE 2.2

Studies and Research Organizations Assessing Global Groundwater Availability

Study/Research Organization	Link
Korzun (1974)	https://unesdoc.unesco.org/ark:/48223/pf0000030551
Shiklomanov and Rodda (2003)	https://unesdoc.unesco.org/ark:/48223/pf0000132004
Döll (2009)	https://iopscience.iop.org/article/10.1088/1748-9326/4/3/035006
Wada et al. (2010)	https://agupubs.onlinelibrary.wiley.com/doi/10.1029/2010gl044571
Richey et al. (2015)	https://agupubs.onlinelibrary.wiley.com/doi/full/10.1002/2015WR017349
IGRAC	https://www.un-igrac.org/
FAO	https://www.fao.org/3/y4502e/y4502e00.htm
WHYMAP	https://www.whymap.org/

Abbreviations: FAO, Food and Agriculture Organization; IGRAC, International Groundwater Resource Assessment Center; WHYMAP, Worldwide Hydrogeological Mapping and Assessment Programme.

Groundwater availability varies significantly around the world. Regions with porous and permeable geological formations, such as sedimentary basins and alluvial plains, tend to have more abundant groundwater resources. In contrast, areas with impermeable geological formations, such as shale, often have less available groundwater (Figure 2.2). Notable regions with large groundwater reserves include the Indo-Gangetic Plain in South Asia, the North China Plain, the High Plains aquifer in the United States, and the Guarani Aquifer System in South America. For example, the Indo-Gangetic Plain aquifer system accounts for about 25% of global groundwater abstraction, making it a crucial water source for drinking and irrigation in South Asian countries, particularly India, Pakistan, Nepal, and Bangladesh (Saha & Sahu, 2015).

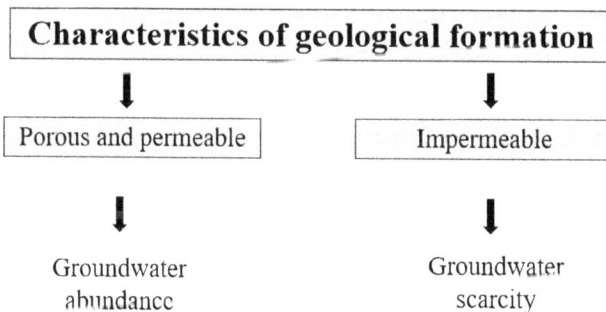

FIGURE 2.2 Influence of geological formation characteristics on groundwater abundance and scarcity.

Over the past five decades, groundwater availability has critically decreased in many regions across the globe. This decline is primarily due to several factors.

First, altered precipitation patterns, coupled with frequent and severe climate events such as droughts and heatwaves, have significantly decreased groundwater recharge rates (i.e., the rate at which groundwater is replenished from precipitation). Numerous studies highlight this trend. For example, McKenna and Sala (2018) found that in the Southwestern USA, for every 1% decrease in annual precipitation, the average playa groundwater recharge rates dropped by 5%. Similarly, in southern Germany, heatwave events led to a reduction in groundwater recharge by up to 26%, compared to a reference period from 1971 to 2000 (Xanke & Liesch, 2022). In China's Xinjiang Uygur Autonomous Region, Qin et al. (2023) showed that due to altered precipitation patterns, groundwater recharge decreased from approximately 150 million m³ in 1980 to around 60 million m³ in 2021.

Second, excessive groundwater pumping, primarily for irrigation, drinking water supply, and industrial purposes, has led to significant depletion of groundwater resources in many areas. Over-pumping causes a range of environmental problems, including water scarcity, saltwater intrusion, soil degradation, and land subsidence. For instance, in Spain, groundwater pumping increased dramatically from 2000 Mm³ per year in 1960 to about 6500 Mm³ per year in 2006, resulting in a sharp decline in the water table, saltwater intrusion, wetland destruction, and serious social conflicts (Ribeiro, 2007). In Mexico City, excessive groundwater extraction has led to the depletion of the city's major aquifers, resulting in severe land subsidence (Sowter et al., 2016).

Third, groundwater pollution has become a major threat to its availability. Pollution can occur when contaminants, such as nitrates from agricultural runoff, industrial spills, and waste disposal, make their way into the groundwater. Nitrates from agricultural activities and urban wastewater are some of the most widespread pollutants globally (Lall et al., 2020). Effective groundwater pollution management requires a combination of regulatory measures, proper waste management, innovative pollution prevention strategies, land use planning, continuous monitoring, and public awareness to encourage responsible resource use.

Fourth, land use and land cover (LULC) changes, such as deforestation, urbanization, and agricultural intensification (e.g., excessive irrigation, heavy fertilizer use, and wetland drainage), can significantly reduce natural water infiltration and the volume of water reaching the groundwater. Studies have demonstrated that such changes can decrease regional groundwater recharge. For example, Siddik et al. (2022) found that urbanization in northwestern Bangladesh reduced regional groundwater recharge by an average of 17 mm per year. In Nebraska National Forest, USA, Adane and Gates (2015) revealed that the transition from grasslands to trees could reduce recharge rates by up to 94%.

Fifth, rapid population growth, with one billion people added every 15 years, increases the demand for water and food. Groundwater is often pumped to meet these rising demands, further depleting aquifers and exacerbating water scarcity in many regions. Regions facing both rapid population growth and increasing water scarcity include South Asia (India, Pakistan, and Bangladesh), Sub-Saharan Africa (Ethiopia, Nigeria, and Kenya), Central Asia (Uzbekistan, Turkmenistan, and Kazakhstan), and the Middle East and North Africa (MENA).

Finally, transboundary aquifers, which span multiple countries, present additional challenges to groundwater availability. Uncoordinated use of groundwater from shared aquifers can lead to overexploitation, water table depletion, and reduced aquifer storage capacity. This can also result in heightened competition, illegal water pumping, and conflicts between bordering countries. Uncertainties surrounding groundwater resources may increase the potential for future conflicts (Fienen & Arshad, 2016).

In conclusion, addressing the issue of groundwater availability requires long-term, comprehensive management strategies that balance water demand with natural recharge rates. These strategies must promote water conservation, prevent pollution, and resolve transboundary groundwater conflicts. Collaboration among policymakers, scientists, stakeholders, and local communities is essential to mitigate the effects of groundwater depletion and ensure water security for future generations (Fienen & Arshad, 2016).

2.3.1.1.3   Glaciers and Ice Caps

The importance of glaciers and ice caps for freshwater storage and ecosystem stability cannot be overstated. They are crucial for regulating water availability in many regions, storing large volumes of freshwater in frozen form, and releasing it gradually over time. However, their rapid shrinking due to rising temperatures poses these significant challenges for water security and ecosystems.

2.3.1.1.3.1   Freshwater Supply Depletion   As glaciers melt, the release of freshwater initially increases, but this volume peaks before declining due to the reduced size of the glacier. This "Peak Meltwater" phenomenon, observed by Huss and Hock (2018), results in a decrease in the long-term contribution of glaciers to freshwater supply. The decline in glacial runoff, especially during dry periods or when glacier melt is minimal, can lead to seasonal water shortages, affecting downstream communities that rely on glacier-fed rivers for their water supply.

2.3.1.1.3.2   Sea-level Rise   Rising global temperatures contribute to both the thermal expansion of seawater and the melting of glaciers, which adds to the overall volume of water in the oceans. The IPCC reported that between 1961 and 2016, glaciers lost around 9000 billion tons of ice, contributing to a 27 mm increase in global mean sea level. This rise leads to more frequent coastal flooding, erosion, and saltwater intrusion into groundwater, which poses risks to coastal ecosystems, infrastructure, and food security. For example, the US Mid-Atlantic region saw the area affected by saltwater intrusion double from 2011 to 2017, making 7600 ha of farmland unsuitable for agriculture (Mondal et al., 2023).

2.3.1.1.3.3   Disruption of Climate Patterns   Glaciers not only respond to temperature increases but also influence regional and global weather patterns. Their shrinking can alter atmospheric circulation and precipitation patterns, resulting in more frequent extreme events such as floods, droughts, and heatwaves. This disruption in climate systems can exacerbate the challenges faced by regions already struggling with water scarcity.

2.3.1.1.3.4   Disruption of Ecosystem Services   Glaciers play a crucial role in maintaining ecosystem stability by regulating the water supply throughout the year, transporting nutrients to aquatic habitats, and shaping landscapes. Their loss can disrupt these functions, leading to significant biodiversity loss and ecosystem instability. Jacobsen et al. (2012) projected that the complete loss of glaciers in the Antisana Ecological Reserve in Ecuador could result in the extinction of up to 38% of endemic species in the area.

2.3.1.1.4 Snowpack

Snowpack is the deposition of snow on the Earth's surface during the winter season, commonly occurring in mountainous areas. Due to its large area, high mountain ranges, and unique atmospheric circulation patterns, Asia has the greatest snow cover extent (SCE) of any region on Earth. More specifically, it accounts for about 50% of the world's total SCE (Young, 2023).

In addition to its primary role in cooling the Earth's climate, snowpack serves as an important freshwater source for snow-fed rivers and streams after it melts, providing freshwater to billions of people. It also contributes to increased soil moisture availability for plants. Furthermore, snowpack can significantly enhance nutrient and enzyme diffusion to plant roots, as it acts as an insulating layer, protecting the soil beneath it from extreme temperature fluctuations. This ensures a gradual and steady supply of nutrients to the soil over time (during its melting) and provides a habitat for soil microbial activity (e.g., bacteria and fungi), which play an essential role in nutrient cycling. However, it is important to note that the positive effects mentioned above strongly depend on the frequency of freeze–thaw cycles. In other words, moderate freeze-thaw cycles can improve soil aeration, nutrient mobilization, and seed germination, boosting plant growth and productivity. However, excessive freeze-thaw cycles can cause physical damage to plant tissues, disrupt root systems, and result in soil erosion, ultimately decreasing plant growth and productivity (Phillips & Nickerson, 2015).

Hydrologists typically consider the snow water equivalent (SWE) to be a key property of snowpack, as it reflects the water content in snow that directly contributes to runoff. SWE is defined as the amount of liquid water that would result from the complete melting of the snowpack per unit area (Eq. 2.1). Accurate measurement of SWE is essential for forecasting spring runoff, evaluating flood risk, and managing water resources.

$$\text{SWE} = \frac{h_s * \rho_s}{\rho_w} \qquad (2.1)$$

where the SWE is the snow water equivalent [cm], h_s is the snow depth [cm], ρ_s is the density of snow [kg m^{-3}], and ρ_w is the density of water [kg m^{-3}].

2.3.1.2 Blue Water Consumption

It is estimated that global average annual blue water consumption ranges from 1000 to 1700 km³ per year (Zhongwei et al., 2019). This water is responsible for irrigating a significant portion of global cropland and is essential for ensuring stable crop yields, particularly in regions with insufficient rainfall. Blue water accounts for approximately 40% of the world's total food production, underscoring its importance in securing global food production and ensuring food security (Steduto et al., 2012; Zisopoulou & Panagoulia, 2021).

With the rapid growth of the global population and the increasing demand for food, blue water consumption in agriculture has risen significantly, largely driven by the expansion of irrigated cropland (Mehta et al., 2024). In regions with limited or highly variable rainfall, reliance on blue water for irrigation becomes even more critical, making it a key factor in meeting the growing food demands of the global population. Mehta et al. (2024) found that about 52% of irrigation expansion has occurred in areas already water-stressed in 2000, such as India, the Southwest United States, Northern China, and Australia. In regions like the Middle East and North Africa, where aquifers are being exploited faster than they can naturally recharge, overreliance on blue water has led to groundwater-level declines, seawater intrusion, wells running dry, and other environmental issues (Jasechko et al., 2024). Jasechko et al. (2024) revealed that groundwater-level declines have accelerated over the past four decades in 30% of the world's regional aquifers. Additionally, increasing temperatures and shifting rainfall patterns have intensified these issues, putting even greater pressure on water resources and making it more difficult to sustain agricultural production in already water-scarce regions.

Ensuring the sustainable use of blue water in agriculture is critical for long-term food security. Approaches such as improving irrigation efficiency, adopting precision irrigation techniques, changing cropping patterns, using alternative water sources (e.g., treated wastewater), and ensuring that increases in irrigated areas occur in regions where water is relatively abundant can help optimize blue water use while maintaining crop productivity (Nouri et al., 2020). Furthermore, policy measures that promote water pricing, encourage aquifer recharge, and implement integrated water resource management will be essential in balancing agricultural water demands with the protection of freshwater ecosystems (Meran et al., 2021).

2.3.2 GREEN WATER

Green water refers to the portion of precipitation that is stored in the soil as moisture and subsequently used by plants and crops via evapotranspiration for growth and development (Johansson et al., 2016). This underscores its crucial role in supporting crop growth, pasturelands, forestry, and terrestrial ecosystems (Hoekstra et al., 2011). On average, about 60% of precipitation is converted into green water (Ahmed et al., 2022), making it the largest freshwater resource that sustains approximately 60% of global food production and feeds 4.8 billion people (He & Rosa, 2023). However, green water can only be used in situ by plants and is vulnerable to fluctuations in temperature and precipitation patterns, which are becoming more erratic due to climate change (He & Rosa, 2023).

Green water security refers to the ability of rainfall and soil moisture to meet crop water requirements, enabling agricultural production without stressing the natural water cycle. GWS occurs when rainfall is insufficient to meet these needs, often worsened by factors such as prolonged droughts, soil degradation, and changing precipitation patterns (He & Rosa, 2023). This scarcity can lead to water-stressed crops (Liu et al., 2022a; He & Rosa, 2023), necessitating the use of supplemental blue water (irrigation) to maintain crop growth and prevent water stress (Pereira et al., 2002). Such vulnerability highlights the importance of adaptive strategies to manage and optimize green water use in rainfed agricultural systems. In other words, improving soil moisture retention and water use efficiency (WUE) through conservation practices (e.g., mulching, reduced tillage, cover crops) is crucial to ensuring food security.

Rosa et al. (2020) estimated that the annual consumption of green water in global croplands is around 5406 km³ per year. They also found that, to avoid reductions in crop yields due to GWS, an additional 2860 km³ per year of blue water would be required globally for irrigation. This blue water, usually sourced from aquifers or surface water bodies, would help maintain crop productivity in regions with insufficient rainfall or other water stress factors (Rosa et al., 2020). These findings stress the need to balance both green and blue water resources to ensure food security under the growing challenges posed by climate change and water scarcity.

Schyns et al. (2015) is one of the most comprehensive studies worldwide that reviewed indicators of green water availability and scarcity. The study serves as an excellent reference for understanding the various indicators (equations, measures, etc.) that can be used to assess green water in agriculture. It examined 80 indicators, highlighting the critical role of green water in water scarcity assessments, alongside the traditionally emphasized blue water resources such as groundwater and surface water.

The study classified these indicators into three main groups:

1. *Green water availability indicators*: They assess the amount of green water available in a given region. However, the study noted that as water demand increases, the values of these indicators may not adequately reflect this change. Within this category, two sub-categories can be distinguished:
   - *Absolute green water availability indicators*: These measure the actual quantity of green water available in absolute terms.
   - *Relative green water availability indicators*: These compare actual green water availability to a reference condition perceived as normal, often defined by long-term climate averages or median values.
2. *GWS indicators*: They assess the extent to which green water availability meets or falls short of demand. They consider both supply and consumption, highlighting regions where green water is insufficient to sustain vegetation or agricultural productivity.
3. *GWS*: This category evaluates GWS in relation to human and ecosystem demands. It accounts for factors such as WUE, land management practices, and climate variability. Indicators in this group help identify areas where green water resources are under pressure due to increasing demand, land degradation, or changing climatic conditions. Overall, agricultural GWS occurs when the amount of rainfall is unable to meet crop water requirements.

2.3.3 DRIVERS OF BLUE AND GREEN WATER INTERACTIONS

In agriculture, green water and blue water are closely interlinked through both natural and anthropogenic processes (Khalili et al., 2023). This means that changes in one often affect the other. When rainfall is sufficient, crops rely on natural mois-ture, which reduces the need for irrigation. Conversely, during dry spells, irrigation (using blue water) becomes essential to meet crops' water needs, highlighting that an increase in one water source typically reduces the dependence on the other (Mao et al., 2020).

The relationship between green and blue water is influenced by various natu-ral and human factors. Key environmental drivers include precipitation, soil type, vegetation, and climate variables such as temperature, humidity, and wind speed, all of which impact evapotranspiration and water availability (Khalili et al., 2023). Soil nutrients and atmospheric CO_2 levels also play a role by affecting plant growth and water uptake (Khalili et al., 2023). Human activities, such as irrigation, dam construction, and land management practices, further alter green water–blue water interactions by modifying runoff, infiltration, and evaporation (Khalili et al., 2023). Additionally, extreme climatic events, such as prolonged droughts and warming trends, can shift the balance between green and blue water, influencing water stor-age and availability over time (Khalili et al., 2023). Understanding these natural and human factors is essential for managing water resources effectively, especially in the face of climate change and increasing water demands. Moreover, water resources assessment approaches should place greater emphasis on these factors (McDonnell, 2017).

By understanding and managing the interplay between green and blue water, farmers and water managers can better plan for both wet and dry periods. This bal-anced approach is especially crucial in regions facing water scarcity, ensuring that agricultural practices remain sustainable and water security is maintained (Mao et al., 2020).

2.3.4 ECONOMIC DIMENSION OF AGRICULTURAL WATER SECURITY

Economic water scarcity (EWS) arises when water resources are physically avail-able, but financial, institutional, or infrastructural limitations restrict access to them (Vallino et al., 2020). This is a critical issue in many regions, such as the Central African region and Central Asian countries, where inadequate investment in water management, irrigation systems, and storage facilities prevents farmers from utiliz-ing available water efficiently.

In some regions, high costs associated with water extraction, distribution, and treatment, combined with poor governance and inequitable water pricing, contribute to EWS. Additionally, increasing water demands from non-agricultural sectors, such as industry, urban development, and energy, can further exacerbate the economic dimension of water scarcity (Gruère et al., 2020). Intensive groundwater extraction for irrigation in semiarid areas also leads to a gradual decline in aquifer levels, rais-ing the costs of water access and intensifying the economic challenges of agricul-tural water security. Moreover, even in water-abundant regions, ineffective water management and governance can result in EWS. Vallino et al. (2020) demonstrated that EWS is not necessarily linked to low physical water availability, challenging common assumptions about the relationship between water scarcity and resource abundance.

To assess water scarcity, the International Water Management Institute (IWMI) developed an indicator that integrates both physical and economic dimensions (Seckler et al., 1998). This indicator measures the proportion of a country's water supply derived from renewable freshwater resources available for human needs while considering existing water infrastructure, such as desalination plants and reservoirs. A key feature of this index is its ability to account for a country's potential to develop water infrastructure and improve irrigation efficiency. More recently, Vallino et al. (2020) highlighted the SDG Indicator on Integrated Water Resources Management (IWRM) as a valuable tool for evaluating EWS in agriculture.

Addressing EWS requires strategic investments in infrastructure, improved governance, and financial mechanisms that ensure equitable and sustainable access to water for agricultural production (Vallino et al., 2020). This includes promoting policies that balance economic and environmental priorities, supporting sustainable agricultural practices, and improving water distribution systems to ensure that water reaches those who need it most (Gruère et al., 2020).

2.3.5 GOVERNANCE DIMENSION OF AGRICULTURAL WATER SECURITY

From a management perspective, water governance determines who gets water, when, and how, as well as their rights to related services and benefits (Allan, 2001). The governance dimension of agricultural water security focuses on how institutions, policies, and regulations manage water for agriculture. Effective water governance is key to achieving water security and sustainable water management (Sismani et al., 2024). It promotes equitable and sustainable water use through transparency, accountability, and participation. It involves frameworks for water allocation, conservation, and pollution control, as well as enforcing regulations to safeguard water rights. Poor governance, such as weak institutions, corruption, and mismanagement, exacerbates water scarcity and hinders access, particularly in areas with limited resources. Countries with strong governance frameworks, like the USA, have implemented efficient systems such as drip irrigation and wastewater reuse, while regions with weak governance, like sub-Saharan Africa, face inefficiencies and infrastructure issues, worsening water scarcity (Gebrehiwot & Gebrewahid, 2016). In this region, the lack of robust water governance may not only lead to physical water scarcity for immediate consumption but also negatively impact agricultural yields (Vallino et al., 2020).

2.4 INDICATORS OF AGRICULTURAL WATER SECURITY

Depending on the objective of the assessment, the spatial extent of the agricultural systems being investigated, and the availability of data, agricultural water security can be assessed at multiple spatial scales, ranging from the farm level to the global scale (Falkenmark, 2013). Assessments at each scale offer valuable insights into the drivers, effects, and potential solutions for enhancing agricultural water security, leading to more sustainable management of water resources. Integrating information across different scales is essential for developing holistic and effective water management strategies that address the diverse needs and challenges of agricultural communities.

The choice of scale for assessment is crucial in determining which indicators are relevant and how they should be interpreted to understand agricultural water security comprehensively. At smaller scales, such as a farm, indicators might focus on factors like soil moisture and irrigation efficiency. In contrast, at larger scales, such as regional or national levels, indices may include factors such as overall water availability, allocation policies, climate change projections, and socioeconomic factors affecting water use in agriculture.

2.4.1   INDICATORS AT THE FARM SCALE

Farm-level water security is achieved when farmers have access to a reliable and sustainable supply of water of adequate quality to meet the needs of their farming activities, such as irrigation and livestock production (Wichelns, 2015). Therefore, assessing agricultural water security at the farm level involves evaluating various indicators that reflect the availability, accessibility, suitability, and sustainability of water resources for farming. The following are the most important indicators.

2.4.1.1   Water Availability

In agriculture, water availability refers to the total volume of water accessible for agricultural use. To assess it at the farm level, several indicators are used, depending on the type of water resource (e.g., precipitation, groundwater well, or surface water) and the farm's hydrological context (such as weather patterns, topography, and soil properties). The following are the most commonly used indicators.

2.4.1.1.1   Rainfall Patterns

Rainfall patterns significantly affect water availability in both rainfed and irrigated agricultural areas. This is because rainfall plays a dual role: it directly enhances green water resources by replenishing soil moisture and indirectly increases the total volume of blue water resources via surface runoff and groundwater recharge processes. Analyzing historical rainfall data is essential to understanding the availability of precipitation as a water source for productive farms. At a specific farm, this analysis typically involves four major steps (Figure 2.3):

Step 1: Data collection
Hourly, daily, monthly, or annual rainfall records should be obtained from a weather station located at or near the farm being investigated. Generally, precipitation is measured using a device known as a rain gauge.

Step 2: Data cleaning
Rain gauges are susceptible to errors due to factors such as wind turbulence, evaporation from collected water, and environmental changes (e.g., an increase in tree cover) (Ahmed, 2018). Therefore, rainfall records must be checked to ensure their accuracy and reliability, reducing the risk of errors in the analysis. Any missing values, duplicate records, or unrealistic measurements in the dataset must be addressed using data-cleaning methods (Ganti & Sarma, 2013).

1. Data collection		2. Data cleaning		3. Data analysis		4. Probability estimation
Hourly, daily, monthly, or annual rainfall records	⇨	Unrealistic missing, and duplicate data must be addressed	⇨	Cleaned data anlalysis through statisitical methods	⇨	Estimation of the likelihood of extreme climate events

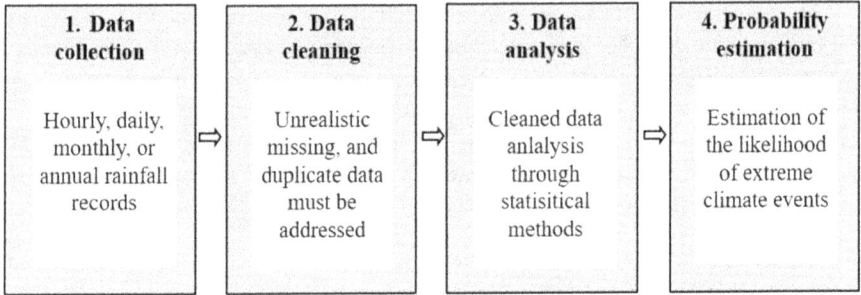

FIGURE 2.3   Steps required for analyzing historical rainfall data.

Step 3: Temporal analysis
Rainfall data over a given period should be analyzed using statistical methods such as linear regression or time-series analysis to detect trends and assess their statistical significance.

Step 4: Analysis of extreme rainfall events
This step is necessary to determine appropriate actions for managing water resources at the farm level and preparing for extreme rainfall events (i.e., floods and droughts).

Rainfall Intensity-Duration-Frequency (IDF) curves are commonly used to understand rainfall extremes in terms of return periods. According to Dupont and Allen (2000), these curves are graphical representations of the probability that a certain rainfall intensity will occur within a given timeframe. Typically, to develop IDF curves, long-term historical rainfall observations must be available, and the following four sequential steps should be followed:

1. Identify extreme rainfall intensities for specific durations using annual maximum analysis.
2. Fit the extreme rainfall intensity time series for each duration to a theoretical distribution function such as Pearson III (Eq. 2.2), Log-Pearson III (Eq. 2.3), or Log-Normal (Eq. 2.4).
3. Calculate rainfall intensity for each duration and return period based on the chosen distribution function.
4. Develop the IDF curves using empirical formulas such as Talbot (Eq. 2.5), Bernard (Eq. 2.6), Kimijima (Eq. 2.7), or Sherman (Eq. 2.8) through regression techniques (Sun et al., 2019).

$$\text{Distribution of Pearson III} \quad P(x) = \frac{1}{\alpha \cdot \Gamma(\beta)} \cdot \left[\frac{x-\gamma}{\alpha} \right]^{\beta-1} \cdot e^{-\left[\frac{x-\gamma}{\alpha} \right]} \quad (2.2)$$

$$\text{Distribution of Log} - \text{Pearson III} \quad P(x) = \frac{1}{\alpha \cdot x \cdot \Gamma(\beta)} \cdot \left[\frac{lnx-\gamma}{\alpha} \right]^{\beta-1} \cdot e^{-\left[\frac{lnx-\gamma}{\alpha} \right]}$$

$$(2.3)$$

$$\text{Distribution of Log} - \text{Normal} \quad P(x) = \frac{1}{(x-a) \cdot \sigma_y \alpha \cdot \sqrt{2\Pi}} \cdot e^{-\frac{\left[\ln(x-a) - \mu_y\right]^2}{2\sigma_y^2}}$$

(2.4)

where α, β, and γ are the three parameters that indicate the scale, shape, and position of the dataset, respectively, and $\Gamma(\beta)$ is the Gamma function.

$$\text{Talbot Equation} \quad I = \frac{a}{d+b} \tag{2.5}$$

$$\text{Bernard Equation} \quad I = \frac{a}{d^2} \tag{2.6}$$

$$\text{Kimijima Equation} \quad I = \frac{a}{d^2+b} \tag{2.7}$$

$$\text{Sherman Equation} \quad I = \frac{a}{(d+b)^e} \tag{2.8}$$

where I is the rainfall duration (mm h^{-1}), d is the duration (h), and a, b, and e are the regression parameters related to the metrological conditions.

By following the above-mentioned steps, researchers can effectively analyze rainfall data to understand local rainfall dynamics and their implications for agricultural water management.

2.4.1.1.2 Well's Specific Capacity

The specific capacity (SC) of a well refers to its efficiency in supplying water to a farm (or farms) over a given period without a significant drop in the water level (Driscoll, 1986). Mathematically, SC is defined as the ratio of the pumping rate (Q) to the water-level decline ($h_0 - h_1$) at the end of a pumping test (Summers, 1972; Todd & Mays, 2005):

$$SC = \frac{Q}{h_0 - h_1} \tag{2.9}$$

where the SC is the specific capacity [L s^{-1} m^{-1}], Q is the pumping rate [L s^{-1}], h_0 is the snow depth [m], h_1 is the water level in the well during pumping [cm], and h_0 is the water level in the well when the pumping is not running [m].

High SC values indicate that the aquifer has sufficient water available to sustain the well's pumping rate without significant declines in water levels, whereas low SC values suggest that the aquifer may be under water stress due to over-pumping (Driscoll, 1986).

SC is not a constant; it can vary significantly over time, influenced by several factors such as pumping duration, pumping rate, constraints affecting groundwater flow within the aquifer, and the influence of nearby pumping wells (Kumar et al., 2016).

By monitoring changes in SC over time through standardized tests, water resource managers can detect trends that may indicate shifts in water availability at the farm level. SC should be monitored at least twice a year, while water levels (h_0 and h_1) should be measured monthly to promptly identify potential well issues. Rehabilitation efforts should be initiated when a well's SC decreases by 25% (Driscoll, 1986).

Finally, while SC values provide valuable insights into well productivity and aquifer properties, their interpretation should be approached with caution due to the limitations and uncertainties associated with SC tests (e.g., limited spatial coverage, short test durations, effects of nearby pumping wells, spatial heterogeneity in aquifer properties, and operational parameters) (Shandilya et al., 2022). A more robust SC test design that accounts for these limitations and uncertainties is therefore essential to provide a comprehensive understanding of aquifer properties and well productivity.

2.4.1.2 Water Quality

Water of suitable quality is essential for ensuring sustainable crop and livestock production. In other words, access to good-quality water is crucial for maintaining soil health, increasing crop yields, and ensuring livestock well-being. Poor water quality, particularly when affected by high salinity, heavy metals, or microbial contamination, can lead to reduced productivity and long-term land degradation. For example, studies have shown that irrigation with high-salinity water can lead to serious soil salinity issues, which, in turn, may reduce crop yields by 10–58%, depending on crop tolerance levels and management practices (Haj-Amor et al., 2022). Similarly, in livestock farming, drinking water quality is critical for maintaining optimal animal health and productivity.

However, in recent years, water resource quality has become increasingly questionable due to environmental pollution, climate change, and intensified human activities. As a result, continuous monitoring of water quality, livestock health, and livestock products is essential to prevent and control diseases effectively (Giri et al., 2020).

Good-quality water is particularly essential in semiarid and arid regions, where both water scarcity and quality pose significant challenges. For instance, in the Middle East, the increasing reliance on treated wastewater and desalinated water for agriculture necessitates rigorous water quality monitoring to prevent soil degradation and maintain crop yields. To mitigate potential risks, an integrated approach is recommended, combining advanced wastewater treatment technologies, community engagement, regulatory oversight, and targeted monitoring (Benaafi et al., 2024). This ensures that treated wastewater used for irrigation does not lead to soil salinization, heavy metal accumulation, or crop contamination, thereby safeguarding both agricultural productivity and environmental sustainability.

Water quality assessment involves analyzing physical, chemical, and biological parameters by collecting and evaluating samples from specific locations (Tyagi et al., 2013). A widely used tool for assessing the suitability of water for agricultural use is the Water Quality Index (WQI), which effectively condenses complex water quality data into a single value ranging from 0 to 100 (Tyagi et al., 2013). This approach is particularly valuable for water management and supply agencies, as it simplifies the interpretation of large datasets and provides an easily understandable measure of water quality (Uddin et al., 2021).

The determination of WQI generally involves four key steps. First, relevant water quality parameters are selected. Second, the collected data are processed, and each parameter's concentration is converted into a dimensionless sub-index. Third, a weighting factor is assigned to each parameter based on its relative importance. Finally, an aggregation function combines the sub-indices and weighting factors to generate a single WQI value. Water quality is then classified into five categories: Excellent (91–100), good (71–90), medium (51–70), low (26–50), and poor (0–25) (Jonnalagadda & Mhere, 2001; Luo et al., 2024).

Numerous WQI models have been developed globally, including the Horton Index, NSF-WQI, OWQI, and CCME-WQI (Rana et al., 2018; Ashwin et al., 2022; Maansi et al., 2022; Marselina et al., 2022). These models differ in their structural design, parameter selection, weighting criteria, and sub-indexing and aggregation methods (Uddin et al., 2021). Such variations influence how different water quality parameters contribute to the final index, affecting the sensitivity and applicability of WQI to specific waterbodies and environmental conditions (Chidiac et al., 2023).

2.4.1.3   Water Use Efficiency

WUE in rainfed or irrigated cropping systems refers to the total plant biomass or crop yield produced per unit of water applied through rainfall and irrigation (Katerji et al., 2008). When considering only irrigation water, it is referred to as irrigation water use efficiency (IWUE), whereas when accounting solely for rainfall, it is termed rainfall water use efficiency (RWUE) (Wichelns, 2015).

Overall, WUE values depend on five key factors:

1. Climate variables, particularly rainfall patterns and atmospheric evaporative demand.
2. Soil properties, which influence water retention and availability.
3. Irrigation water management, including scheduling and efficiency.
4. Agronomic practices, such as crop rotation, tillage, and fertilization.
5. Genotypic characteristics of crops, including yield potential, drought tolerance, and maturity period (Ierna, 2023).

Given these influencing factors, it is crucial to consider them collectively when interpreting RWUE values.

Estimates of WUE provide valuable insights into how effectively plants utilize water to produce crop yields or biomass, aiding efforts to enhance agricultural productivity in both rainfed and irrigated systems (Wichelns, 2015). Typically, IWUE and RWUE are calculated using the following formulae (Eq. 2.10 and Eq. 2.11, respectively) (Katerji et al., 2008):

$$IWUE = \frac{Y_{gi} - Y_{gd}}{I} \qquad (2.10)$$

where Y_{gi} is the crop yield harvested during a growing season through irrigation (kg ha^{-1}), Y_{gd} is the dryland yield, i.e., the crop yield without irrigation, (kg ha^{-1}) (often, in semiarid and arid areas, Y_{gd} may be 0), I is the total amount of water applied during the same growing season (mm), and RWUE is expressed in kg ha^{-1} mm^{-1}.

$$RWUE = \frac{Y_{gd}}{R} \qquad (2.11)$$

where Y_{gd} is the dryland yield during a growing season (kg ha^{-1}), R is the total rainfall during the same growing season (mm), and RWUE is expressed in kg ha^{-1} mm^{-1}. Efficient use of irrigation water and rainfall is crucial for achieving optimal crop yield (Morell et al., 2011). WUE at the field level can be enhanced through various soil and water conservation measures that optimize water management, reduce crop evapotranspiration, and minimize water losses (Wichelns, 2015). The most effective strategies include:

1. Efficient irrigation methods, such as drip irrigation, to reduce water wastage.
2. Water-saving technologies, including automated irrigation scheduling systems, to optimize water application.
3. Conservation tillage, such as no-till or reduced tillage, to improve soil moisture retention.
4. Crop selection, prioritizing varieties that utilize water more efficiently.
5. Proper soil management practices, including mulching, cover cropping, and organic matter addition, to enhance water retention.
6. Proper fertilizer management, ensuring nutrient availability without excessive water use.
7. Improved drainage, preventing waterlogging and salinity buildup.

Enhancing WUE at the field level provides multiple agronomic and environmental benefits, such as:

- Cultivating a larger area using the same volume of irrigation water or rainfall.
- Mitigating water shortages, particularly in water-scarce regions.
- Saving energy by reducing the need for excessive pumping and irrigation.
- Improving soil health, enhancing long-term agricultural sustainability.

2.4.1.4 Water Application Efficiency (E_a)

Irrigation water application efficiency (E_a) provides valuable insights into how effectively an irrigation system delivers water to the crop root zone, minimizing losses due to runoff, evaporation, and deep percolation beyond the root zone (Figure 2.4). In other words, E_a represents the ratio of water effectively used for crop growth to the total water applied.

Mathematically, E_a is commonly calculated using the formula developed by Burt et al. (1997) (Eq. 2.12):

$$Ea = \frac{V_s}{V_f} \qquad (2.12)$$

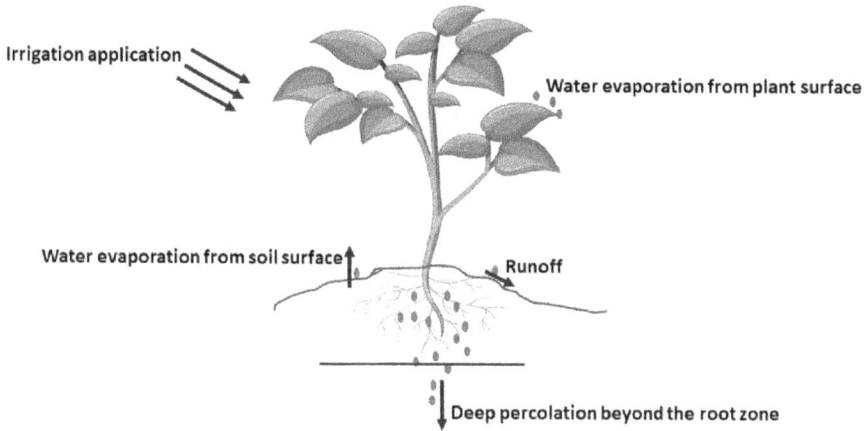

FIGURE 2.4 Water losses after irrigation application.

where V_s is the volume of irrigation water retained in the root-zone (mm), V_f is the total volume of irrigation water delivered to the plot (mm), and E_a is expressed in %.

Literature-based reference values for irrigation water application efficiency (E_a) typically fall within a narrow range, influenced by several factors, including irrigation system type, soil texture, water application rate, crop variety, weather conditions, water management practices, and water quality.

Table 2.3 provides a detailed summary of these factors and their impact on E_a. By carefully considering and managing these elements, farmers can enhance E_a, optimize water use, and improve crop productivity while minimizing irrigation water waste.

TABLE 2.3
Main Factors Affecting Water Application Efficiency (E_a)

Factor	Specific Factor	Typical Range of E_a
Irrigation system	Surface irrigation	45 (conventional) to 80% (precision level basin)[*]
	Sprinkler irrigation	65 (hand move) to 90% (LEPA system)[*]
	Micro-irrigation	80 (bubbler) to 95% (subsurface drip irrigation)[*]
Soil texture	Fine-texture	Higher E_a values
	Coarse-texture	Lower E_a values
Water application	Quick water application	Lower E_a values
	Slow water application	Lower E_a values
Crop varieties	Climate-adapted crops	Higher E_a values
Weather	Cooler temperatures	Higher E_a values
Water management	Water-saving measures	Higher E_a values
Water quality	High water salinity	Lower E_a values

[*]*Data source:* Irmak S. et al. 2011; data for well-designed and well-managed irrigation systems.

2.4.1.5 Crop Water Productivity

CWP is a key indicator for optimizing agricultural water use, as it reflects how efficiently water is converted into crop production (Molden et al., 2010). It is calculated by dividing the total crop yield (kg) by the total volume of water used for irrigation or transpired by the crop (m³). A higher CWP indicates better WUE (Zwart & Bastiaanssen, 2004).

CWP is relevant at multiple scales, from individual farms to regional and national levels, as it varies significantly based on factors such as crop type, soil properties, and irrigation management (Mabhaudhi et al., 2021). It is particularly useful for comparing irrigation methods and guiding optimal irrigation strategies (Mpakairi et al., 2024).

Moreover, monitoring and analyzing CWP during extreme climate events can help identify crops that are most vulnerable to water stress, supporting the development of climate-resilient agricultural strategies (e.g., Chowdhuri et al., 2022; Mpakairi et al., 2024). However, while CWP is a valuable indicator, it should be considered alongside water accessibility, affordability, and long-term sustainability to provide a comprehensive assessment of agricultural water security.

2.4.2 Indicators at Basin Level

At the basin level, agricultural water security can be comprehensively assessed through the integration of hydrological, ecological, and management concepts. These concepts help evaluate both water availability and consumption, while also accounting for ecosystem interactions and human interventions.

Recently, the WF concept has gained significant attention as a robust indicator for assessing water security at this scale. Unlike conventional water-use assessments, the WF quantifies not only the total water withdrawn for agricultural activities but also the return flows, water that re-enters the watershed after use, offering a more holistic perspective on water consumption.

This approach enhances our understanding of water allocation, identifies inefficiencies, and supports the development of sustainable water management strategies that maintain both agricultural productivity and ecological balance (Veettil et al., 2022).

The WF is typically divided into three components: Blue, green, and grey WF. The blue WF refers to the volume of irrigation water consumed in crop production and not returned to the same basin (Hoekstra et al., 2011). The green WF represents the portion of rainwater stored in soil or vegetation that evaporates or transpires during plant growth (Falkenmark, 2003). The grey WF is the volume of freshwater required to dilute pollutants to acceptable water quality standards (Franke et al., 2013).

WF accounting offers a detailed view of how water is used in a river basin, which is helpful to guide improvements in water resources management, especially in basins already experiencing water stress. This accounting is often performed according to these steps.

2.4.2.1 Data Collection

This step involves gathering hydrological data (e.g., precipitation, streamflow), agricultural water use (e.g., irrigation volumes, crop water demands), industrial and domestic water withdrawals, and return flow data.

2.4.2.2   WF Calculation

The blue WF and green WF are computed using water balance models such as CROPWAT, AquaCrop, and CropSyst. The grey WF is estimated using field data on agricultural practices, crop yields, and natural background pollution levels, combined with models like SWAT (Soil and Water Assessment Tool) or MODFLOW to quantify pollutant loads entering rivers or aquifers (D'Ambrosio et al., 2019).

2.4.2.3   Spatial and Temporal Assessment

This step examines how WF varies across different locations and seasons within the basin, helping to identify regions of high-water stress and seasonal fluctuations in water availability.

2.4.2.4   Sustainability Analysis

WF values are compared with available renewable water resources to assess water stress levels. Indicators such as the Agricultural Water Stress Index (AWSI) (e.g., Xinchun et al., 2017; Rezaei et al., 2019) are used. The AWSI evaluates the extent to which both green and blue water availability can meet agricultural water requirements. It directly indicates water scarcity in both rainfed and irrigated areas, which the blue water scarcity index (BWSI), defined as the ratio of the blue water footprint (BWF) to BWA, alone cannot fully capture (Liu et al., 2022b). Mathematically, AWSI is defined as the ratio of the total crop WF to the availability of agricultural water resources over a given period (Eq. 2.13) (Xinchun et al., 2017).

$$AWS = \frac{AWF}{AWR} \tag{2.13}$$

where AWF is the total crop WF (the volume of blue and green water used for agricultural production) over a given period (m^3), and AWR is the available agricultural water resources during the same period (m^3).

An AWSI value close to 1 indicates a high level of water stress and potential unsustainability, as agricultural water consumption nearly equals the available resources. In contrast, lower AWSI values suggest that water use remains within sustainable limits, ensuring sufficient resources to meet agricultural demands (Xinchun et al., 2017). A persistently high AWSI may highlight the need for improved water management strategies, such as optimizing irrigation practices, adopting water-efficient technologies, or implementing policies to enhance water resource sustainability.

Mekonnen and Hoekstra (2012) conducted the first global assessment of BWS across 405 major river basins from 1996 to 2005 using the BWSI. Their study revealed that two-thirds of these basins experienced severe water scarcity for at least one month each year, with critical hotspots identified in India, China, the western USA, the Middle East, and North Africa. In the Indus and Ganges basins, the BWSI often exceeded 200%, indicating that water consumption more than doubled the available renewable supply. The study emphasized the importance of monthly assessments, as many basins that appear sustainable on an annual scale may still face severe shortages during dry months.

More recently, Liu et al. (2022b) assessed the evolution of agricultural water scarcity across 228 major river basins worldwide from 1971 to 2010 using the AWSI. Their findings indicated a clear trend of increasing water scarcity severity over the study period, particularly in key agricultural regions. They also demonstrated that 22% of river basins were classified as water-scarce based on the AWSI, compared to only 8% using the BWSI.

2.4.3 INDICATORS AT NATIONAL LEVEL

Agricultural water security is a key national concern that requires effective indicators for assessment. Composite indicators help evaluate water availability (e.g., total renewable water resources per capita, agricultural water withdrawal, groundwater depletion), efficiency (e.g., irrigation efficiency, water productivity, distribution losses), governance (e.g., national water policies, Water Governance Index, infrastructure investment), and sustainability (e.g., drought frequency, soil moisture deficit, water quality) (Bahramifard & Zibaei, 2024).

Multi-Attribute Decision-Making (MADM) methods, such as the Analytic Hierarchy Process (AHP), Technique for Order Preference by Similarity to Ideal Solution (TOPSIS), and Weighted Sum Model (WSM), provide structured evaluations for policymaking. AHP prioritizes indicators by assigning weights based on national needs, such as water availability in arid regions or irrigation efficiency in agricultural areas. TOPSIS ranks regions by comparing criteria such as water productivity and climate risks, identifying best-performing areas. WSM calculates an overall water security index by weighting key factors, guiding national strategies for sustainable water use. By integrating MADM methods, policymakers can make data-driven decisions to enhance agricultural water security, improve water management, and ensure long-term food sustainability (Zolghadr-Asli et al., 2021).

2.4.4 INDICATORS AT GLOBAL LEVEL

Agricultural water security is a critical global challenge that demands comprehensive assessment and sustainable management of water resources. Effective evaluation requires consideration of key factors such as water availability, quality, management practices, and climate resilience. By integrating these aspects, global agricultural water security can be more effectively assessed, and strategies can be implemented to mitigate risks and enhance sustainability.

Various indicators, including GWS, BWS, and EWS, are used to assess agricultural water security. Rosa et al. (2020) mapped these indicators across global croplands for 130 primary crops in a spatially explicit and integrated manner, showing that water scarcity varies by location and season. Their findings indicate that 76% of global croplands experience GWS for at least one month annually, while 42% face it for five months each year. Expanding their analysis to Agricultural Economic Water Scarcity (AEWS), they found that 15% of croplands (0.14 billion hectares) suffer from AEWS, and 16% are irrigated unsustainably. AEWS is particularly prevalent in low-income countries with significant yield gaps, largely due to inadequate investments in irrigation infrastructure. In contrast, high-income and arid regions have fewer EWS croplands, as irrigation expansion has been effective in boosting food and feed production. Notably, two-thirds of AEWS croplands are concentrated in Sub-Saharan Africa, Eastern Europe, and Central Asia.

Multiple global assessments of water scarcity (e.g., Liu et al., 2022b; Rosa et al., 2019; Mekonnen & Hoekstra, 2016) have consistently highlighted those regions in the middle to low latitudes of the Northern Hemisphere face severe water scarcity. Additionally, both physical and economic water stress contribute to widespread water scarcity across nearly all African countries. This is largely due to limited economic capacity to develop water infrastructure and persistent poverty, which hinder access to available water resources, even in areas where water is physically abundant. Furthermore, nearly half of the world's irrigated agriculture and approximately 4 billion people live in regions experiencing water scarcity for at least one month each year (Mekonnen & Hoekstra, 2016; Rosa et al., 2019).

2.5 LIMITATIONS

Indicators used to assess water security in agriculture face several limitations, primarily related to data availability, incomplete datasets, spatial and temporal variability, measurement errors, and oversimplification.

First, water security varies significantly across locations and seasons, but indicators often provide averaged values that fail to capture the localized effects of extreme events, such as droughts and floods. This highlights the need for further research into the spatial analysis of drought and flood vulnerability assessments to better understand localized impacts on communities and infrastructure (Dhawale et al., 2024). It is also essential to examine the spatial variability in the occurrence, intensity, and impacts of these events to improve water resource allocation (Dhawale et al., 2024).

Second, Schyns et al. (2015) revealed that operationalizing GWS indicators still faces challenges. For example, determining when and where green water can be productively used is complex, requiring both absolute and relative indicators, alongside crop growth models to assess availability. Additionally, estimating green water consumption in forestry is difficult due to trees' deep roots accessing both green and blue water.

Furthermore, many indicators fail to account for water quality issues, such as salinity and pollution, which can significantly impact agricultural productivity. Water availability and quality are closely interconnected, both playing a crucial role in agricultural production and food security. Therefore, assessments of agricultural water security should be comprehensive, accounting for the dependencies between availability and quality. Achieving this requires the incorporation of water quality and ecological parameters into water availability assessment models, which can only be accomplished through sustained collaboration between experts in water availability modelling, water quality modelling, and ecological processes (e.g., van Vliet et al., 2024).

Another major limitation is the lack of socioeconomic and policy considerations in many water security indicators. While these indicators often focus heavily on hydrological and climatic factors, they frequently overlook governance, economic conditions, and equity in water distribution. Factors such as conflicts over water access, pricing policies, and institutional management play a crucial role in agricultural water security but are not always integrated into assessments.

Finally, as climate change disrupts precipitation and evapotranspiration patterns, it further exacerbates water scarcity. Therefore, a more integrated approach that accounts for future uncertainties in rainfall patterns, temperature shifts, and groundwater depletion should be adopted when assessing agricultural water security. This approach will help ensure more accurate predictions of climate change risks and support effective water management strategies (Rosa & Sangiorgio, 2025).

2.6   CONCLUSIONS

Agricultural water security plays a vital role in socioeconomic development, agricultural production, and food security. Over the last four decades, it has gained significant political and economic attention. To mitigate the impacts of diminishing water resources, a better understanding of the relationship between water demand and supply is essential. This study provides an overview of agricultural water security assessments, focusing on indicators across various scales, from farm-level management to global policy frameworks.

At the farm level, indicators such as irrigation efficiency and water productivity help optimize water use while maintaining crop yields. At the basin level, indicators such as blue and green WFs influence long-term agricultural sustainability. At the national level, governance plays a crucial role, with indicators like water withdrawal percentages, policy frameworks, and investment in irrigation infrastructure shaping water security strategies. At the global level, agricultural water security can be assessed using various indices, such as GWS, BWS, EWS, and AEWS.

Strengthening data collection, improving governance, and adopting adaptive water management strategies are essential for ensuring long-term sustainability. By aligning practices at the farm level with strategies at the basin, national, and global levels, policymakers and stakeholders can build resilient agricultural systems that can withstand future water challenges.

3 Contribution of Hydrological Modeling to the Assessment of Agricultural Water Security under Climate Change
Case Studies from Around the World

3.1 INTRODUCTION

Ensuring water security for agriculture, particularly for irrigation and livestock production, has become a critical global challenge and one of the most pressing issues of the 21st century (Aligholi & Hayati, 2022). This challenge arises from multiple interrelated factors. Firstly, population growth has driven the expansion of irrigated areas and increased agricultural water demand to meet rising food production needs. At the same time, urbanization and industrial development have placed additional pressure on existing water supplies (Fukase & Martin, 2020; Mancuso et al., 2020). Secondly, beyond water scarcity in many regions, industrial growth has polluted previously clean water resources, degrading water quality and reducing their usability (Liu et al., 2016; Mancuso et al., 2020). Moreover, the over-extraction of groundwater and excessive surface water withdrawals have led to significant resource depletion, undermining long-term water availability. Wada et al. (2010) reported a sharp increase in groundwater depletion, from 126 km³ year⁻¹ in 1960 to 283 km³ year⁻¹ in 2000.

Furthermore, climate change has exacerbated hydrological variability, resulting in erratic rainfall patterns, prolonged droughts, and an increase in extreme weather events. An analysis by Konapala et al. (2020), covering the period from 2005 to 2100, revealed a significant increase in precipitation variability across 35.6% of the Earth's surface. These compounding factors are worsening the imbalance between water supply and demand, particularly in the agricultural sector (Fiorillo et al., 2021).

This imbalance can further strain water storage capacities, intensifying the ongoing water crisis in many countries (La Jeunesse et al., 2016; Fiorillo et al., 2021). It

DOI: 10.1201/9781003660521-3

may also lead to higher water costs, reduced crop yields, and increased regional food insecurity. Fitton et al. (2019) estimated that approximately 11% of current croplands and 10% of grasslands globally are vulnerable to declining water availability, potentially leading to reduced agricultural productivity, especially in Africa, the Middle East, China, Europe, and Asia.

As such, understanding how climate change affects agricultural water supply and demand is essential for developing effective long-term adaptation strategies (Parandvash & Chang, 2016; Fiorillo et al., 2021). Globally, numerous studies have examined the impacts of climate change on agricultural water security, focusing on regional variations in water demand (e.g., Masia et al., 2018; Fiorillo et al., 2021) and water supply (e.g., Peres et al., 2019; Tarekegn et al., 2022). In these studies, advanced hydrological models have played a key role in assessing climate change effects on hydrological processes and supporting decision-making in water resource management. Each model possesses specific characteristics, strengths, and uncertainties. In this context, this chapter provides a comprehensive review of modeling approaches used to assess the impacts of climate change on agricultural water security, critically examining their capabilities, limitations, and associated uncertainties.

3.2   OVERVIEW OF HYDROLOGICAL MODELS

3.2.1   MODEL DESCRIPTION

A hydrological model is a simplified representation of a real-world hydrological system, designed to enhance the understanding, forecasting, and management of water resources (Sahu et al., 2023). These models serve as essential tools in various applications, including water resource management, flood forecasting and control, drought assessment, and soil degradation assessment (Table 3.1).

They integrate diverse datasets, including climatic variables (e.g., rainfall, air temperature), soil characteristics (e.g., texture), topography, vegetation cover, hydrological observations, and other physical parameters, to simulate the movement, distribution, and availability of water resources (Devia et al., 2015).

More specifically, a hydrological model functions as an input–output system that simulates changes in water storage and fluxes, as well as potentially related chemical

---

TABLE 3.1

Applications of Hydrological Modeling: Purposes and Case Studies

Application	Purpose	Case Study
Water resource management	Optimize water allocation for various uses	Dutta and Sarma (2021)
Flood forecasting	Predict flood and support mitigation strategies	Venkata Rao et al. (2024)
Drought assessment	Evaluate water scarcity and support conservation	Tenagashaw and Andualem (2022)
Climate change studies	Assess climate change effects on hydrological processes	Xu et al. (2005)
Soil degradation assessment	Evaluate land degradation risks (e.g., erosion)	Almasalmeh et al. (2022)

and physical properties, across the Earth's surface and subsurface. This is typically achieved through the water balance equation, which is based on the principle of mass conservation (Horton et al., 2022). The model accounts for the horizontal movement of water across terrain via channels, vertical exchanges between the atmosphere, surface, and groundwater, and fluctuations in water storage within a defined control volume.

In general, the water balance is expressed by the following equation (Eq. 3.1) (Sutcliffe, 2004; Mohajerani et al., 2021):

$$P + I = ET + Q + \Delta S + \Delta G + \Delta W \qquad (3.1)$$

where P is precipitation, I is water inflow, ET is evapotranspiration, Q is water outflow, and ΔS, ΔG, and ΔW represent changes in soil moisture, groundwater storage (saturated zone), and surface water storage, respectively.

At the plot and catchment scales (Figures 3.1 and 3.2), the water balance is represented by Equations 3.2 and 3.3, respectively (Zhang et al., 2002; Mohajerani et al., 2021):

$$\Delta S = P - I - E - T - Q - DP + CR \qquad (3.2)$$

where ΔS is the change in water stored in soil, P is precipitation, I is interception, E is soil evaporation, T is plant transpiration, Q is runoff, DP is deep percolation to groundwater, and CR is capillary rise.

$$\Delta S + \Delta G + \Delta W = P - ET - Q \qquad (3.3)$$

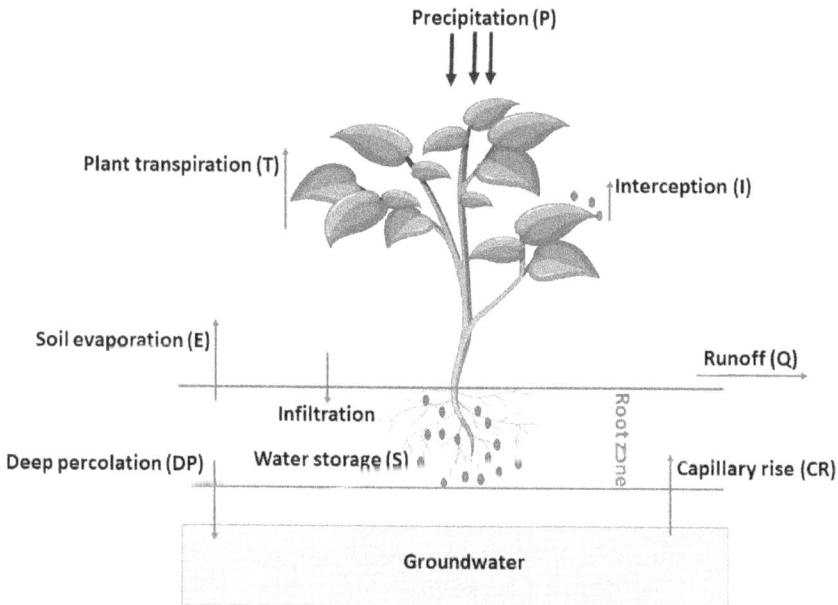

FIGURE 3.1   Main components of water balance at the plot scale.

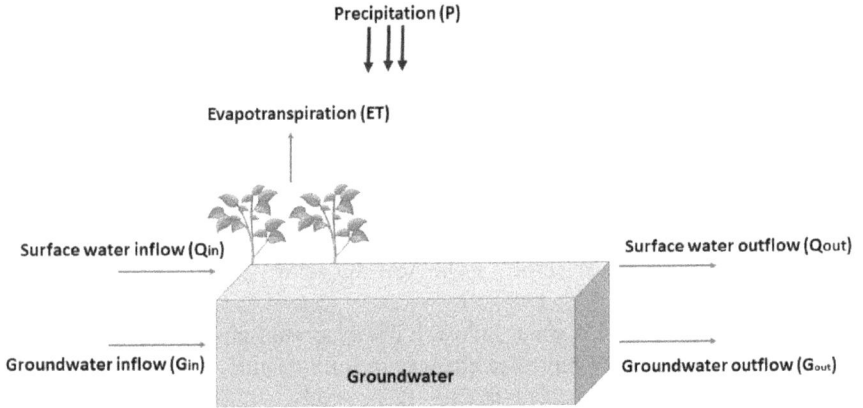

FIGURE 3.2   Main components of water balance at the catchment scale.

where ΔS, ΔG, and ΔW are the changes in water stored in soil, groundwater, and surface water bodies, P is precipitation, ET is evapotranspiration, and Q is water flow out of the watershed.

At the global scale, the water balance reflects the exchange of water among the land, ocean, and atmosphere. Resolving the water balance at this level requires global observations of key components of the hydrological cycle, e.g., precipitation, evapotranspiration, and changes in water storage (Mohajerani et al., 2021). Thanks to advancements in observational techniques, satellite sensors, and modeling tools, such global assessments have become increasingly feasible.

According to Shiklomanov (2004), the Earth's total annual precipitation, which is numerically equivalent to total evaporation, is approximately 577,000 km³, or 1130 mm. The evaporation depth from the ocean surface is about three times greater than that from land. Each year, around 87% of global evapotranspiration occurs over the oceans, with only 13% originating from terrestrial surfaces.

3.2.2   MODEL TYPES

The main types of models used in hydrological modeling include the following.

3.2.2.1   Conceptual Models

Conceptual hydrological models are simplified representations of hydrological processes that require relatively fewer data inputs. Typically designed for precipitation-runoff simulations, these models rely on observed or assumed empirical relationships among various hydrological variables (Liu et al., 2017b). They use simple mathematical equations to represent key processes such as evapotranspiration, surface storage, percolation, snowmelt, baseflow, and runoff, rather than solving complex partial differential equations (PDEs). This simplification reduces computational demands but necessitates the use of multiple parameters, some of which may not have direct physical meaning (AghaKouchak & Habib, 2010).

Conceptual models are generally lumped, meaning that they apply uniform parameter values across an entire watershed, often ignoring spatial variability (Jaiswal et al., 2020). They depend heavily on observed data, and their accuracy is closely tied to the quality of these inputs (Jaiswal et al., 2020). While they are computationally efficient and data-light compared to other model types, making them suitable for regional studies, they tend to oversimplify complex hydrological processes.

Over the past few decades, many conceptual models have been developed. The Stanford Watershed Model IV (SWM), introduced by Crawford and Linsley in 1966, was the first major conceptual model and featured 16 to 20 parameters (Crawford & Linsley, 1966). Other widely used models include HEC-1 (Burnash et al., 1973), Boughton (1984), HYMOD (Moore, 1985), Modhydrolog (Chiew & McMahon, 1994), HBV (Bergström, 1995), the Xinanjiang model (Zhao & Liu, 1995), and the ARNO model (Todini, 1996).

A review by Devia et al. (2015) highlighted four key characteristics of conceptual models: (1) Parameters are derived from field data and typically require calibration; (2) They are relatively simple and can be easily implemented in computer code; (3) Large datasets of hydrological and meteorological variables are often required; and (4) Calibration frequently relies on curve fitting, which can make physical interpretation difficult.

3.2.2.2 Physically-based Models

Physically-based models simulate hydrological processes using fundamental physical laws, offering high accuracy and the ability to represent real-world processes in a detailed manner. However, they require extensive input data and large datasets, which are not always readily available. Compared to conceptual models, physically-based models are more complex and computationally demanding, as they involve the numerical solution of PDEs. Their application often demands significant effort in model calibration and expertise in both hydrology and numerical methods (Thirumalaiah & Deo, 2000; Kim et al., 2015).

The development and implementation of physically-based models also require high-quality, site-specific data and a deep understanding of hydrological systems. Data scarcity and variability can pose substantial challenges, particularly for short-term predictions, where errors may arise due to local input limitations (Costabile & Macchione, 2015). Furthermore, uncertainties in input data, such as soil characteristics, precipitation, and land use, can affect model accuracy and limit predictive reliability (Özdoğan-Sarıkoç & Dadaser-Celik, 2024).

One of the most widely used physically-based hydrological models is the Soil and Water Assessment Tool (SWAT), developed by the Agricultural Research Service (ARS) of the United States Department of Agriculture (USDA) (Arnold et al., 1998). SWAT is designed to simulate hydrological processes at the watershed scale over long-term periods, up to 100 years (Li et al., 2018). It has been extensively applied in various research domains, including the assessment of climate change impacts on water resources (Sood et al., 2013), water quality evaluations (Pisinaras et al., 2010), and forecasting of runoff, sediment transport, and nutrient loading (Zewde et al., 2024). Moreover, SWAT has been used to analyze the effects of land use, land cover change, and agricultural practices (Marhaento et al., 2017).

SWAT conceptualizes runoff generation and watershed hydrology through two interconnected components: The land surface and the water surface (Li et al., 2018).

- The land surface component, also known as the slope hydrological process, governs runoff formation and slope flow dynamics within sub-basins. It simulates the movement of water, sediments, and nutrients, incorporating eight essential modules: Hydrology, meteorology, sediment transport, soil temperature, crop growth, nutrient cycling, pesticide dynamics, and agricultural practices (Li et al., 2018).
- The water surface component, referred to as the channel hydrological process, manages the routing and redistribution of water, sediments, and dissolved substances, such as nitrogen, phosphorus, and pesticides, through river networks (Li et al., 2018). This component includes the dynamics of flow in channels, reservoirs, ponds, and wetlands, governed by channel and water body flow equations (Kannan et al., 2007; Li et al., 2018).

The interactions between these components are governed by the model's water balance equation, expressed as follows (Eq. 3.4):

$$\text{SW}_t = \text{SW}_0 + \sum_{i=1}^{t} \left(\text{R}_{day} - \text{R}_{surf} - \text{E}_a - \text{W}_{deep} - \text{Q}_{gw} \right) \tag{3.4}$$

where SW_t represents the final soil water content, SW_0 denotes the initial soil water content, t is the time (days), and R_{day} is the daily precipitation amount (mm). Q_{surf}, E_a, W_{deep}, and Q_{gw} refer to surface runoff, surface evaporation, the amount of water percolating into the aquifer, and the return flow, respectively. All of these parameters are measured on day i and expressed in millimeters (mm).

The SWAT model is constructed through several key steps. First, the study area is defined, and the necessary input data are collected, including the digital elevation model (DEM), climate, soil, land use, and management data. The model is then set up using a GIS-based interface (e.g., ArcSWAT or QSWAT), during which the watershed is delineated, hydrologic response units (HRUs) are defined, and weather data are integrated. Once the setup is complete, the model is executed, and any errors are identified and addressed.

Subsequently, calibration and validation are carried out either manually or with the aid of tools such as SWAT-CUP. This process involves adjusting model parameters based on observed data to ensure alignment between simulated and measured results. Finally, the performance of the model is assessed using various statistical indicators, including the coefficient of determination (R^2, Eq. 3.5), Nash–Sutcliffe efficiency (NSE, Eq. 3.6), and error indices such as Mean Absolute Error (MAE, Eq. 3.7) and Root Mean Square Error (RMSE, Eq. 3.8):

$$R^2 = \left(\frac{\sum_{i=1}^{n} \left(O_i - \bar{O} \right) \left(S_i - \bar{S} \right)}{\sqrt{\sum_{i=1}^{n} \left(O_i - \bar{O} \right)^2} \sqrt{\sum_{i=1}^{n} \left(S_i - \bar{S} \right)^2}} \right)^2 \tag{3.5}$$

$$NSE = 1 - \frac{\sum\limits_{i=1}^{n}(O_i - S_i)^2}{\sum\limits_{i=1}^{n}(O_i - \bar{O})^2} \tag{3.6}$$

$$MAE = \frac{\sum\limits_{i=1}^{n}|(S_i - O_i)|}{n} \tag{3.7}$$

$$RMSE = \sqrt{\frac{1}{n}\sum\limits_{i=1}^{n}(S_i - O_i)^2} \tag{3.8}$$

where S_i and O_i are the ith simulated and observed values, respectively, $\bar{O}$ is the average of observed values, n is the number of observations and is the average of observed values.

In addition to the SWAT model, other widely used physically based models include the Precipitation-Runoff Modeling System (PRMS), MIKE SHE, the Variable Infiltration Capacity (VIC) model, the Hydrologic Modeling System (HEC-HMS), and MODFLOW (Sahu & Chandniha, 2025). Further details about these models are presented in Table 3.2.

TABLE 3.2
Commonly Used Physically-Based Hydrological Models

Model	Developer	Main Purpose	References
Soil and Water Assessment Tool (SWAT)	USDA Agricultural Research Service (ARS)	Predict the impact of land management practices on water, sediment, and nutrient yield in watersheds	Arnold et al. (1998)
Precipitation-Runoff Modeling System (PRMS)	The United States Geological Survey (USGS)	Assess the impact of different climate and land use combinations on streamflow	Markstrom et al. (2015)
MIKE SHE	DHI Water & Environment	Simulating surface water flow and groundwater flow	DHI (2005)
Variable Infiltration Capacity (VIC)	Xu Liang, University of Washington (UW), USA	Evaluate hydrological responses to climate variability and land surface interactions	Liang et al. (1994)
Hydrologic Modeling System (HEC-HMS)	United States Army Corps of Engineers (USACE)	Simulate the precipitation-runoff processes of dendritic watershed systems	Sahu et al. (2023)
MODFLOW	United States Geological Survey (USGS)	Simulate groundwater flow and interactions with surface water	Harbaugh (2005)

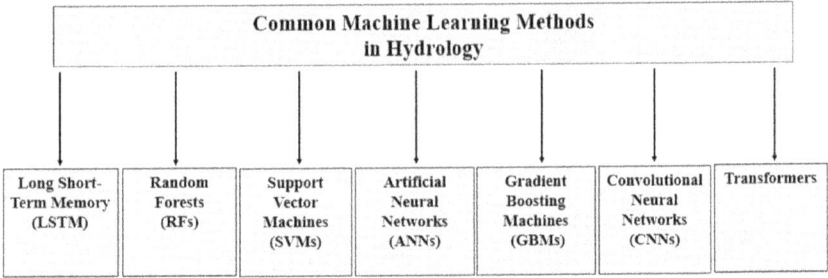

```
┌─────────────────────────────────────────────────────────────────────┐
│                   Common Machine Learning Methods                      │
│                            in Hydrology                                │
└─────────────────────────────────────────────────────────────────────┘
```

| Long Short-Term Memory (LSTM) | Random Forests (RFs) | Support Vector Machines (SVMs) | Artificial Neural Networks (ANNs) | Gradient Boosting Machines (GBMs) | Convolutional Neural Networks (CNNs) | Transformers |

FIGURE 3.3 Common machine learning methods in hydrology.

3.2.2.3 Machine Learning Models

Machine learning (ML) models are data-driven approaches that leverage historical patterns to predict hydrological variables such as streamflow, precipitation, groundwater recharge, and flood occurrence. These models are capable of identifying complex non-linear relationships and offer rapid predictions. However, they typically lack the ability to explain the underlying physical processes and require large, high-quality training datasets to perform effectively.

Hasan et al. (2024) reviewed machine learning applications in hydrology and identified seven key methods (Figure 3.3):

1. *Long Short-Term Memory* (*LSTM*): These models are designed to capture temporal dependencies in hydrological time series, enhancing the simulation of streamflow, rainfall-runoff, and groundwater levels compared to traditional methods. Their primary limitations include high computational demands and a tendency to overfit the training data.

2. *Random Forests* (*RFs*): RFs utilize an ensemble of decision trees to improve prediction accuracy. They are particularly effective for flood mapping, drought forecasting, and precipitation prediction. Although they handle noisy data well, they may introduce bias when applied to small datasets.

3. *Support Vector Machines* (*SVMs*): SVMs construct optimal hyperplanes for classification and regression tasks. They have proven effective in streamflow forecasting, soil water content estimation, and groundwater level prediction. While SVMs perform well in high-dimensional feature spaces, they can be sensitive to noisy input data.

4. *Artificial Neural Networks* (*ANNs*): ANNs are well-suited to model complex non-linear relationships in hydrological systems. They are widely applied in rainfall-runoff modeling and flood prediction. However, they are prone to overfitting, which may limit their ability to generalize to unseen data.

5. *Gradient Boosting Machines* (*GBMs*): GBMs build decision trees iteratively to minimize error, improving the prediction of flood events, soil moisture dynamics, and groundwater levels. Despite their high accuracy, they require careful hyperparameter tuning to mitigate overfitting.

6. *Convolutional Neural Networks* (*CNNs*): Originally developed for image processing, CNNs are increasingly used in hydrology for spatial data analysis, particularly in remote sensing applications, precipitation estimation, and flood mapping. They efficiently process large datasets but typically require substantial labelled training data.

7. *Transformers*: Adapted from Natural Language Processing (NLP), Transformers are well-suited for sequential hydrological data. They have demonstrated strong performance in streamflow forecasting and flood prediction due to their ability to model long-range dependencies in time series data.

3.3   HYDROLOGICAL MODELING STEPS

Hydrological modeling involves simulating the movement, distribution, and quality of water across the hydrological cycle. It supports understanding of the processes within a hydrologic system, aids in predicting responses to defined management scenarios, and contributes to the development of sustainable water resources strategies (Baran-Gurgul & Rutkowska, 2024). Although the modeling steps may vary depending on the scale, purpose, and tools applied, Ochoa–Tocachi et al. (2022) proposed the following ten essential steps.

3.3.1   POLICY QUESTION DEFINITION

This step focuses on specifying the core questions the model aims to address (e.g., water excess, water scarcity), defining the spatio-temporal scale, expected outputs (e.g., maps, data, indicators), and the required level of accuracy. It is essential to identify the decision-making context to ensure that the model produces outputs aligned with stakeholder needs. The primary objective is to translate these policy questions into a modeling framework that delivers relevant, precise, and actionable results (Ochoa–Tocachi et al., 2022).

3.3.2   IDENTIFICATION OF HYDROLOGICAL ECOSYSTEM SERVICES

Here, the key hydrological ecosystem services, such as water provision, baseflow maintenance, flood attenuation, or sediment retention, are identified based on the model's purpose. For instance, flood attenuation would be prioritized when assessing flood risks (e.g., Terêncio et al., 2020). This step ensures that the chosen services are adequately represented in the modeling process (Ochoa–Tocachi et al., 2022).

3.3.3   SPECIFICATION OF NATURAL INFRASTRUCTURE INTERVENTIONS

In this step, natural infrastructure interventions (e.g., reforestation, river restoration) are defined and assessed for compatibility with the selected hydrological model. Key variables such as land use, soil water storage, and surface features are integrated into the simulation to evaluate impacts on hydrological responses. Adjustments to parameters like vegetation cover, infiltration rates, and river morphology help assess the effectiveness of interventions in managing water flows (Ochoa–Tocachi et al., 2022; Dixon et al., 2016).

3.3.4   ASSESSMENT OF DECISION-MAKING CONTEXT

This involves evaluating the institutional, technical, and socioeconomic environment within which the modeling will take place. Considerations include available human capacity, budget constraints, data availability, and the tools/software that can be utilized (Ochoa–Tocachi et al., 2022).

3.3.5   HYDROLOGICAL MODEL SELECTION

An appropriate hydrological model is selected based on the defined objectives and the resolution (spatial and temporal) required. For example, seasonal flow analysis in climate impact studies may focus on seasonal volumes, while flood early warning systems require high temporal resolution and real-time performance (Byaruhanga et al., 2024).

3.3.6   DATA COLLECTION AND PREPARATION

At this stage, both primary (field-based) and secondary (existing) datasets are acquired and processed. Key data include meteorological variables (e.g., rainfall, temperature), DEMs, land use/land cover (LULC), streamflow, groundwater levels, evapotranspiration, sediment loads, water quality, and soil properties (Ochoa–Tocachi et al., 2022). Data sources (Table 3.3) range from ground-based stations to remote sensing tools (Tauro et al., 2018).

Recent advances in web-based platforms have streamlined the acquisition of input datasets. For example, historical climate data can be accessed from the Climate Forecast System Reanalysis (CFSR) (Saha et al., 2010); DEMs from the Shuttle Radar Topography Mission (SRTM) (Farr et al., 2007); and LULC from satellite imagery or agencies like the European Space Agency (ESA) (Jiang & Yu, 2019).

3.3.7   SCENARIO DEVELOPMENT

Scenarios are created to represent various future conditions, land management strategies, or intervention alternatives. They are essential for evaluating how different natural infrastructure interventions affect hydrological systems (Ochoa–Tocachi et al., 2022).

3.3.8   MODEL CALIBRATION AND VALIDATION

Calibration refers to the adjustment of model parameters to enhance the agreement between simulated and observed data, playing a crucial role in minimizing uncertainty in model outputs. In contrast, validation involves evaluating the model's predictive performance using independent datasets that were not utilized during the calibration phase. Together, these processes ensure the model's credibility and robustness. Both graphical and numerical techniques can be employed to assess calibration and validation quality, with the choice of method depending on the nature of

TABLE 3.3
Measurement and Estimation Methods for Input Data Used in Hydrological Models

Input Data	Measurement/Estimation Methods
Rainfall	• Raingauges
	• Radar data
	• Crowdsourced observations
Temperature	• Weather stations
	• Remote sensing
	• Climate reanalysis datasets
Streamflow	• Stream gauging station
	• Acoustic Doppler Current Profilers (ADCP)
River bathymetry	• Echo sounding
	• Remote sensing green LiDAR bathymetry
Soil moisture	• Time Domain Reflectometry (TDR)
	• Remote sensing
	• Gravimetric methods
Evapotranspiration	• Lysimeters
	• Eddy covariance method
	• FAO Penman-Monteith method
Vegetation dynamic	• Time series of satellite imagery collected in the visible and near-infrared wavelengths
Groundwater level	• Observation wells
	• Piezometers
	• Satellite-based methods
Sediment loads	• Turbidity sensors
	• Remote sensing
Water quality	• Laboratory analysis

the simulated variables, the modeling objectives, and whether the model is deterministic or probabilistic (Biondi et al., 2012).

Graphical methods provide intuitive visual assessments of model performance and remain essential, even with the widespread availability of quantitative metrics (Biondi et al., 2012). Common graphical tools include hydrographs and flow duration curves, such as percent exceedance probability plots (Moriasi et al., 2007). Other visualization techniques like bar charts and box plots may also be applied (Moriasi et al., 2007). Hydrographs illustrate the temporal variation of observed and simulated flows, allowing for the identification of model biases and discrepancies in peak magnitude, timing, and recession characteristics. Flow duration curves assess the model's ability to replicate the statistical distribution of daily flows across the calibration and validation periods (Figure 3.4).

Numerical methods rely on quantitative performance metrics to objectively evaluate model accuracy through error analysis. In hydrological modeling, commonly used statistical indices include the coefficient of determination (R^2, Eq. 3.5), NSE (Eq. 3.6), and error-based metrics such as MAE (Eq. 3.7) and RMSE (Eq. 3.8). These

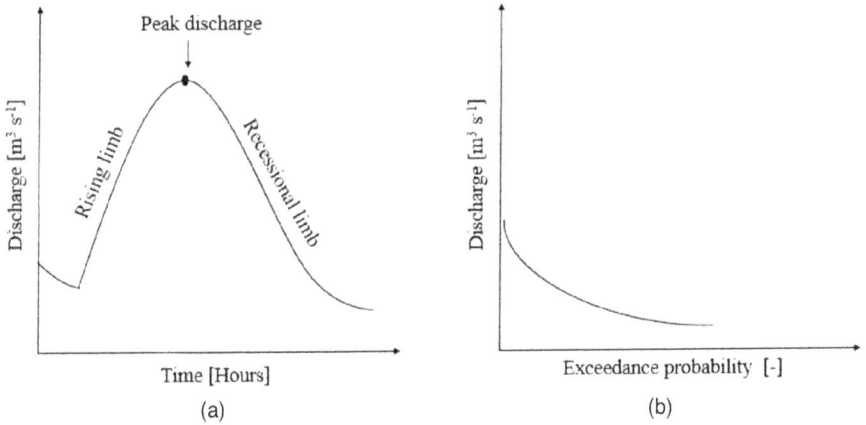

FIGURE 3.4 Examples of a hydrograph (a) and a flow duration curve (b).

indices quantify the discrepancies between observed and simulated values, expressed in either absolute or squared units of the modeled variable, thereby facilitating the interpretation of model performance. An ideal model would yield a value of zero for RMSE, MAE, and Mean Squared Error (MSE), indicating a perfect match between observed and simulated data (Moriasi et al., 2007).

3.3.9 Sensitivity and Uncertainty Analysis

Uncertainty in hydrological modeling arises primarily from errors in input data (e.g., precipitation, temperature), inaccuracies in measured data (e.g., streamflow, groundwater levels), suboptimal parameter values, and structural limitations of the model itself (Refsgaard & Storm, 1990). These uncertainties can considerably impact the credibility of model predictions. To address this, sensitivity analysis (SA) is typically conducted following model setup to assess how variations in model parameters affect model outputs (Herrera et al., 2022). The main objectives of SA are to: (1) Rank parameters based on their influence, (2) exclude parameters with negligible impact, and (3) identify regions in the parameter space that yield extreme model responses (Herrera et al., 2022). When applied effectively, SA aids model calibration by pinpointing key parameters and deepening understanding of model behavior under uncertainty.

SA techniques are generally categorized into local and global approaches (Santos et al., 2022). Local sensitivity analysis (LSA) evaluates the effect of small perturbations in a single parameter at a time, while holding others constant. This approach does not account for parameter interactions. In contrast, global sensitivity analysis (GSA) examines the entire parameter space and considers the combined effects and interactions of all parameters across a broad range of values (Abbas et al., 2024).

Common LSA techniques include the one-at-a-time (OAT) method and differential analysis (DA) (Devak & Dhanya, 2017). These methods typically rely on calculating partial derivatives of model outputs with respect to individual parameters.

A sensitivity index S can be computed for a small perturbation Δe_i in parameter e_i, while all other parameters are held constant, using the following equation (Melching & Yoon, 1996):

$$S = \frac{\dfrac{Y\left(e_1,\ldots\ldots,e_i+\Delta e_i,\ldots\ldots,e_p\right)-Y\left(e_1,\ldots\ldots,e_i,\ldots\ldots,e_p\right)}{Y\left(e_1,\ldots\ldots,e_i,\ldots\ldots,e_p\right)}}{\dfrac{\Delta e_i}{e_i}} \tag{3.9}$$

where Y is the model output, e_i is the model parameter, and Δe_i is the small change applied to parameter e_i.

GSA does not depend on specific input values; instead, it treats the model $f(x)$ as a "black box." Accordingly, global sensitivity indices are designed to evaluate the overall response behavior of the model rather than the accuracy of any individual solution (Sobol, 2001). GSA is therefore particularly valuable for understanding the influence of parameter variability across the entire input space. Among widely used GSA methods, the Morris method (Morris, 1991) provides an effective compromise between computational efficiency and the ability to detect non-linear effects and parameter interactions. The Sobol method (Sobol, 2001), on the other hand, offers a more rigorous and comprehensive decomposition of output variance, enabling detailed quantification of both main effects and higher-order interactions among parameters.

3.3.10   RESULTS INTERPRETATION

Model outputs must be translated into meaningful indicators that align with the specific policy questions and decision-making context. These indicators, such as changes in peak flows, annual water yield, or groundwater recharge, serve as crucial tools for evaluating the impacts of different management scenarios. By linking scientific outputs to real-world concerns, such interpretation supports evidence-based planning, resource allocation, and the formulation of effective policies (Ochoa–Tocachi et al., 2022).

3.4   CASE STUDIES AND APPLICATIONS

In recent years, a wide range of case studies from Europe, Asia, the Americas, Africa, and Australia have demonstrated the practical application of hydrological modeling tools in assessing agricultural water security under climate change and informing adaptive management strategies. Among these tools, the SWAT has emerged as one of the most extensively applied models. Its capacity to simulate complex watershed processes, such as streamflow, sediment transport, nutrient dynamics, and soil water balance, across diverse climatic and geographical contexts has contributed to its broad appeal. SWAT's comprehensive framework allows for detailed, spatially and temporally explicit simulations, making it particularly well-suited for analyzing the impacts of climate variability, land use change, and agricultural practices on water resources. The model's versatility and robustness are evidenced by its application in nearly 4000 published studies worldwide (CARD, 2019; Leong Tan et al., 2020), reflecting its value in supporting water management decisions across multiple basin scales and environmental conditions

3.4.1   CASE STUDIES FROM EUROPE

Hydrological modeling has been successfully implemented in various European countries (Table 3.4) to assess agricultural water security in the context of climate change. The following examples illustrate this successful application, showcasing a range of climatic conditions, farming systems, and adaptation strategies across the continent.

3.4.1.1   Norway

The SWIM model was applied by Huang et al. (2025) to simulate river discharge and hydrological components in six forest-dominated catchments under RCP2.6 and RCP4.5 climate change scenarios for the near future (2041–2070) and far future (2071–2100), relative to the historical period (1976–2005). The simulations projected changes in annual streamflow ranging from −2% to +8%, primarily driven

---

TABLE 3.4

Overview of Case Studies Assessing the Impacts of Climate Change on Agricultural Water Resources in Selected European Countries

Country	Study Area	Hydrological Model (s)	References
Norway	Forest-catchments	SWIM	Huang et al. (2025)
Greece	Aracthos River Basin	SWAT IAHRIS	López-Ballesteros et al. (2020)
Germany	Rur catchment	SWAT	Eingrüber and Korres (2022)
Poland	Narew basin	SWAT WaterGAP	Piniewski et al. (2014)
Austria	Seewinkel	System dynamics model	Valencia Cotera et al. (2023)
Spain	Cidacos River	SWAT	Oduor et al. (2023)
Italy	Aterno-Pescara River	SWAT+	Tariq et al. (2024)
United Kingdom	Five Welsh catchments	SWAT	Dallison et al. (2022)
Netherlands	Drentsche Aa catchment	SIMGRO	Querner et al. (2022)
Sweden	Three small watersheds	SWAT	Grusson et al. (2021)
Belgium	Zenne watershed	SWAT	Leta and Bauwens (2018)
Hungary	Zala River Basin	Budyko-type model	Csáki et al. (2020)
Switzerland	Broye catchment	SWAT	Zarrineh et al. (2020)
France	Garonne River	SWAT	Grusson et al. (2018)
Serbia	Toplica River catchment	HBV-light ANN	Idrizovic et al. (2020)

by climate. Additionally, forest growth was found to reduce climate-induced flood levels by up to 3% in all catchments, except for one micro-catchment affected by extensive clear-cutting.

3.4.1.2 Greece

In northwestern Greece, López-Ballesteros et al. (2020) applied the SWAT model and IAHRIS software to assess the impacts of climate change on the hydrological regime of the Aracthos River Basin. They first demonstrated that the SWAT model was an effective tool for simulating streamflow in the basin, showing satisfactory performance during both the calibration and validation periods. Using five global climate models under RCP4.5 (moderate emissions) and RCP8.5 (high emissions) scenarios for the period 2070–2099, the study projected reductions in precipitation and river flow, along with increases in both maximum and minimum temperatures, relative to the historical period (1970–1999). Furthermore, IAHRIS results indicated significant alterations in the flow regime, particularly with longer and more severe droughts. These changes could have adverse effects on the river's ecosystem, reducing biodiversity, altering seasonality, and weakening both hydraulic and environmental resilience.

3.4.1.3 Germany

A study by Eingrüber and Korres (2022) assessed future trends in extreme precipitation and flooding in the Rur catchment, a tributary of the Maas River, one of Europe's major rivers, using coupled climate-hydrological modeling. The study applied future climate projections under A2a (high emissions) and B2a (moderate emissions) scenarios to predict daily precipitation through 2099. These projections were then used in the SWAT model to simulate runoff and flood patterns. Compared to the 1961–1990 baseline, extreme precipitation in the highlands of the catchment was expected to increase significantly by 33–51%, while trends in the lowlands remained largely insignificant. Runoff simulations indicated that flood magnitudes were projected to rise by 31% (B2a) to 36% (A2a) by 2099.

3.4.1.4 Poland

A study by Piniewski et al. (2014), using the SWAT and WaterGAP models, indicated that climate change poses a risk to the environmental flow regime in the Narew basin. Disruptions to this regime, such as changes in the quantity and timing of water flows, could harm the essential functions and services provided by the basin's ecosystems.

3.4.1.5 Austria

In Seewinkel, a region known for its extensive agriculture, Valencia Cotera et al. (2023) developed a novel hydrological model based on system dynamics to assess climate change adaptation measures. The model was calibrated using hydrological and climate data from the reference period of 1981–2011. They found that under all climate change scenarios, from low to high emissions, combining improved irrigation efficiency with crop changes could reduce water demand by 23% to 40%, thereby enhancing groundwater recharge and supporting sustainable water management.

Specifically, they showed that the local aquifer's level could rise above the historical average by 0.43 m through combined measures: 0.20 m from increasing irrigation efficiency, 0.20 m from changing crops, and 0.06 m from artificial aquifer recharge.

3.4.1.6 Spain

Oduor et al. (2023) used the SWAT model to evaluate future streamflow and nitrate load in the Cidacos River watershed (Navarre) under RCP4.5 (moderate emissions) and RCP8.5 (high emissions) climate change scenarios across short-term (2011–2040), medium-term (2041–2070), and long-term (2071–2100) periods. The study projected consistent declines in both indicators, with the most severe reductions under the long-term RCP8.5, particularly in autumn and winter. These trends were attributed to decreased rainfall and rising temperatures. The authors emphasized the need for adaptive agricultural practices, such as improved irrigation and fertilizer management, to protect water resources and reduce pollution.

3.4.1.7 Italy

In Central Italy, where water resources are vulnerable to Mediterranean climate change, Tariq et al. (2024) assessed the long-term impacts of climate change (2015–2100) on river runoff in the Aterno-Pescara River watershed using the SWAT+ model. Their findings indicated a general increase in daily temperatures (up to 0.6°C per decade) and a decline in precipitation (−16.4 mm per decade), resulting in a negative trend in river runoff (−0.036 m³ s⁻¹ per decade). The study also forecasted more intense and prolonged drought events over the coming decades, underscoring the importance of implementing strategies to mitigate the effects of climate change on water resources in order to secure long-term sustainability.

3.4.1.8 United Kingdom

Dallison et al. (2022) used the SWAT model to estimate future streamflow and water quality variables (nitrogen, phosphorus, suspended sediment, and dissolved oxygen) in five Welsh catchments under the RCP8.5 (high emissions) climate change scenario for the period 2021–2080. Their projections indicated a decline in annual average flows (−4% to −13%), alongside more pronounced seasonal variations, including a 41% increase in spring flows and a 52% decrease in autumn flows. They anticipated further degradation of water quality, as well as more frequent and intense high-flow events in spring and increased low flows in autumn. They also expected that these changes in streamflow and water quality could result in more unreliable, seasonal, and polluted water resources, underscoring the urgent need for adaptive water management strategies to address the growing variability in both flow and water quality.

3.4.1.9 Netherlands

To assess the effectiveness of water management strategies, such as adjusting water levels and managing groundwater abstraction, Querner et al. (2022) used the SIMGRO model to simulate the impact of future climate scenarios on groundwater, surface water, and soil moisture dynamics in the Drentsche Aa catchment. The simulation results emphasized the importance of adaptive strategies for ensuring sustainable water availability for agriculture and natural areas under climate change, highlighting the need to balance ecological requirements with evolving water demands and mitigate the effects of droughts.

3.4.1.10   Sweden

Grusson et al. (2021) used the SWAT model and climate projection techniques to examine the impacts of RCP4.5 (moderate emissions) and RCP8.5 (high emissions) climate change scenarios on hydrological processes in three small agricultural watersheds. The study specifically focused on how changes in precipitation patterns, especially increased rainfall intensity, influence runoff, soil water content, and evapotranspiration. The results indicated that heavier rainfall events, particularly those exceeding 15 mm day^{-1}, led to higher runoff but did not improve soil water content. When precipitation decreased, soil water content also diminished. The most notable impacts were observed during the peak cropping season (May–August), suggesting that although rainfall intensity in southern Sweden is expected to increase, it may not enhance soil water availability for crops, despite the rise in heavy rainfall events.

3.4.1.11   Belgium

In the Zenne watershed, where farming is particularly intensive in the upstream area, Leta and Bauwens (2018) assessed the future impacts of climate change on peak and low flows under four scenarios, wet summer, wet winter, mean, and dry, using the SWAT model. Simulations for the 2050s and 2080s, relative to 1961–1990 data, revealed significant differences across scenarios. Under the wet summer scenario, peak flows increased by up to 109%, raising flood risks and disrupting ecosystems. In contrast, the dry scenario projected reductions of up to 169% in both peak and low flows, intensifying drought, crop losses, and water demand. The most pronounced low-flow declines occurred downstream, highlighting spatial variability linked to groundwater contributions. Overall, the study projected earlier and more severe droughts by 2095, stressing the urgency of timely adaptation strategies.

3.4.1.12   Hungary

A spatially distributed, long-term Budyko-type climate-runoff model, developed specifically for the country, was applied by Csáki et al. (2020) to predict the impact of climate change on actual evapotranspiration and runoff in the Zala River Basin, a major source of inflow to Lake Balaton, the largest shallow lake in Central Europe. The prediction results indicated a slight increase in evapotranspiration, while runoff is projected to decrease significantly, by more than 40% by 2071–2100 compared to the baseline period (1981–2010). This substantial reduction in runoff may hinder Lake Balaton's ability to sustain its current outflow, potentially transforming it into a closed lake system.

3.4.1.13   Switzerland

In Western Switzerland, where climate change is anticipated to affect ecosystem services, Zarrineh et al. (2020) assessed these effects in the Broye catchment using the SWAT model in combination with EURO-CORDEX climate projections for three timeframes: 1986–2015 (baseline), 2028–2057 (near future), and 2070–2099 (far future). The authors reported a significant shift in seasonal water dynamics, with summer low flows expected to decline by 77%, while winter flows may increase by 65% in the far future. A reduction in summer precipitation is projected to lower nitrate leaching by 25%, but due to reduced dilution, nitrate concentrations could rise by 14%. Conversely, increased winter precipitation may lead to a 44% rise in nitrate leaching and an 11% increase in nitrate concentrations, despite higher runoff levels. Overall, these expected changes may threaten sustainable crop production and water management, highlighting the need for improved storage, efficient irrigation, and better nutrient management practices.

3.4.1.14   France

In the southwest of France, Grusson et al. (2018) examined the impact of climate change on the availability of blue and green water in the Garonne River using the SWAT model. Future hydrological trends for 2010–2050 were compared with historical data from the 1962–2010 period. The study revealed a marked decline in soil water content and an increase in winter evapotranspiration (ET), while summer ET decreased in some areas due to limited soil moisture. Blue water (BW) availability also declined significantly in summer, whereas winter river discharge increased in the upper parts of the watershed due to reduced snow accumulation and a rise in liquid precipitation, leading to enhanced surface runoff.

3.4.1.15   Serbia

In southern Serbia, using the Toplica River catchment as a case study, Idrizovic et al. (2020) evaluated the potential effects of climate change scenarios RCP4.5 and RCP8.5 on the hydrological regime over the period 2021–2100, focusing on runoff and the link between groundwater levels and river discharge. Runoff was simulated with a calibrated HBV-light model, and ANNs were used to model groundwater–discharge interactions. Statistical analyses (Mann-Kendall and Mann-Whitney tests) revealed no major annual trends but indicated growing monthly variability. Under RCP8.5, end-of-century winters were projected to experience increased precipitation, temperature, and discharge, leading to higher groundwater levels and flood risk. Conversely, summers showed declining discharge and groundwater levels, posing risks to water supply, agriculture, energy production, and ecosystems.

3.4.2   Case Studies from Asia

In Asia, unpredictable water availability, exacerbated by land use changes and climate variability, poses significant risks to both water and food security. Hydrological modeling plays a crucial role in understanding these risks and developing strategies for sustainable water management. The following case studies (Table 3.5) highlight how such models are applied to assess the impacts of climate change and climate variables on agricultural water security across the Asian continent.

3.4.2.1   Bangladesh

In the Upper Meghna River Basin, the SWAT model was applied by Mamoon et al. (2024) to assess the future hydrological responses of the Barak, Meghalaya, and

TABLE 3.5
Overview of Case Studies Assessing the Impacts of Climate Change on Agricultural Water Resources in Selected Asian Countries

Country	Study Area	Hydrological Model(s)	References
Bangladesh	Meghna River Basin	SWAT	Mamoon et al. (2024)
Pakistan	Ravi River	SWAT	Ullah et al. (2024)
Vietnam	Dak B'la watershed	SWAT	Loi et al. (2020)
China	National (General)	GCAM, Xanthos	Sun et al. (2023)
India	Sind River Basin	SWAT	Narsimlu et al. (2013)
Japan	Shogawa River Basin	HYDREEMS	Ohba et al. (2025)
Malaysia	Langat River Basin	SWAT	Amirabadizadeh et al. (2017)
Indonesia	Central Kalimantan	MLR, RF, XGBoost	Hikouei et al. (2023)
Jordan	Yarmouk River Basin	SWAT	Hammouri et al. (2017)
Nepal	Tamakoshi River Basin	HBV-light	Budhathoki et al. (2023)
Russia	Dvina/Kolyma/ Indigirka Rivers	SWAT	Nasonova et al. (2018)
Turkey	Alata River Basin	HYPE	Yildirim et al. (2021)
Qatar	Qatar (General)	Monte Carlo	Baalousha (2016)
Saudi Arabia	Jazan Province	SWAT	Masoud et al. (2024)
Kuwait	Dammam Formation Aquifer	MODFLOW–MT3DMS	Al-Huwaishel et al. (2022)
South Korea	Anyang watershed	SWAT-MODFLOW	Ware et al. (2023)
Thailand	Southern basin	Machine learning, GR2M	Ditthakit et al. (2021)
Taiwan	Choushui River	SWAT-MODFLOW	Ngo et al. (2024)

Tripura sub-basins under the SSP2-4.5 and SSP5-8.5 climate change scenarios from 2026 to 2100, relative to the historical period 1985–2014. The study found a consistent increase in annual maximum flows, particularly in the Meghalaya sub-basin. Dry season flows were projected to rise moderately (up to 31–50%), while wet season flows could increase more substantially (up to 47–66%) by 2100. Additionally, the projections suggested an increase in the frequency and intensity of extreme flood events, driven by rising mean flows under future climate scenarios.

3.4.2.2   Pakistan

Ullah et al. (2024) studied the combined effects of land use and climate change on the water flow system of the Ravi River using the SWAT model, integrated with GIS and remote sensing. Future land use projections, generated using TerrSet modeling, indicated a 31.7% increase in built-up areas from 2020 to 2100. Climate change scenarios (SSP2 and SSP5), downscaled with the CMhyd model, forecasted increases in precipitation (10.9–14.9%) and temperature (12.2–15.9%) by 2100, relative to 2020. SWAT simulations revealed a projected rise in river inflows (19.4–28.4%) from 2016 to 2100, along with significant shifts in seasonal flow patterns. Historical data also indicated a decline in groundwater depth of 0.8 m per year from 1996 to 2020, largely driven by land use and climate change. The findings underscored the importance of integrated land and water resource management in the face of evolving climate and land use conditions.

3.4.2.3   Vietnam

In the Dak B'la watershed in the Central Highlands of Vietnam, the SWAT model was successfully applied by Loi et al. (2020) to compare water discharge values under baseline climate conditions (2001–2018) with future projections under the A1B climate change scenario for the period 2020–2069. The results indicated a slight increase (0.55%) in average annual water discharge. However, more significant impacts were observed during the flood season, particularly in November 2050, when peak flows (584 m³ s⁻¹) exceeded the historic 2009 flood peak (450 m³ s⁻¹), underscoring the increased flood risk associated with climate change.

3.4.2.4   China

In China, the world's most populous country, where rising greenhouse gas emissions are expected to exacerbate water scarcity, Sun et al. (2023) employed an integrated approach by coupling the Global Change Analysis Model (GCAM) with the Xanthos global hydrological model to project future water footprint and water stress under a range of climate and socioeconomic scenarios, including SSP2–RCP6.0. Water stress was assessed using the Water Stress Intensity (WSI), defined as the ratio of water footprint to renewable water resources. Results suggested that China's water footprint will peak around 2030 before gradually declining. Emission reduction efforts, particularly in the power sector, are expected to significantly shape water demand, with the electricity sector projected to become the largest water user as China transitions to low-carbon energy, potentially worsening scarcity. The most critical period for water stress is projected between 2025 and 2035, especially in northern river basins, but all climate models consistently indicate a decline in water stress after 2050, suggesting long-term improvement.

3.4.2.5   India

Narsimlu et al. (2013) assessed the impacts of future climate change on water resources in the Upper Sind River Basin (India) using the SWAT model. Simulated monthly streamflow showed strong agreement with observed data during calibration and validation periods, confirming the model's reliability. Under the IPCC A1B climate change scenario, and relative to the 1961–1990 baseline, average annual streamflow is projected to increase by about 16% by 2050 and nearly 93% by 2100. However, the findings also indicate a substantial rise in monsoon flows and a decline during non-monsoon months, pointing to potential seasonal imbalances in future water availability. In the same basin, a more recent study by Kumar et al. (2021) applied the SCS-CN method to estimate surface runoff from rainfall over the period 2005–2014, reporting an average annual runoff of 133.71 mm. This corresponds to a total average runoff volume of about 35×10^8 m³, accounting for about 17% of the total average annual rainfall.

3.4.2.6   Japan

Ohba et al. (2025) employed the HYDrological Evaluation with Evapotranspiration Modeling System (HYDREEMS) simulations alongside climate projections to evaluate the effects of climate change on hydrological droughts in the Shogawa River Basin, central Japan. Their findings revealed a reduction in summer streamflow and a significant increase in the number of consecutive drought days. By applying self-organizing maps (SOMs), they linked changes in river discharge to distinct climate and weather patterns, highlighting that drought severity varies depending on these conditions. In particular, future scenarios featuring southerly and easterly airflows are projected to experience more frequent monthly scale droughts, driven by reduced precipitation and heightened evapotranspiration.

3.4.2.7 Malaysia

Amirabadizadeh et al. (2017) assessed the impacts of future climate variables on the hydrological regime of the Langat River Basin using the SWAT model. Future climate variables were downscaled using the Statistical DownScaling Model (SDSM). Projections for the 2030s and 2080s indicated increases in both mean monthly precipitation and maximum temperature. The SWAT model, calibrated using the SUFI-2 algorithm, projected that average annual streamflow could rise by approximately 71% by 2050 and 108% by 2100 relative to the 1990–2011 baseline, accompanied by a notable increase in surface runoff.

3.4.2.8 Indonesia

In Central Kalimantan, Indonesia, Hikouei et al. (2023) modeled groundwater levels within a peat dome affected by drainage canals and recurring wildfires over 2010–2012 using a multilinear regression model (MLR) alongside two machine learning algorithms: RF and extreme gradient boosting (XGBoost). While all models demonstrated strong performance, spatial analysis showed that XGBoost provided superior accuracy near drainage canals, areas prone to fire ignition in peatlands. The study also identified elevation and precipitation as the most influential factors driving spatiotemporal variations in groundwater levels.

3.4.2.9 Jordan

In the Yarmouk River Basin (northern Jordan), an area already facing acute water stress due to ongoing climate change, Hammouri et al. (2017) used the SWAT model, driven by projections from three global climate models, to assess future water availability. Their findings suggested that streamflow could decline by up to 22% by 2080, with the sharpest reductions (35–40%) projected for peak flow months such as February and March. While minor increases in flow may occur in some months, the overall trend indicates increasing water scarcity. The study also found that precipitation is the primary driver of surface runoff vulnerability, whereas temperature plays a comparatively limited role in altering runoff volumes.

3.4.2.10 Nepal

Budhathoki et al. (2023) applied the HBV-light model to the glacierized Tamakoshi River Basin to quantify key water balance components under conditions of limited hydrometric data. The simulation results indicated that streamflow was primarily governed by monsoon rainfall (62%), followed by baseflow (20%), glacier melt (13%), and snowmelt (5%). Peak discharge occurred in August, driven by intense monsoonal rainfall and glacier melt, while the monsoon season showed the greatest fluctuations in water storage. In contrast, the post-monsoon period experienced minimal changes in storage, with baseflow contributions peaking in October and reaching their lowest point in February. The study further revealed that between 2011 and 2020, runoff contributions to streamflow increased, whereas changes in water storage declined. These findings highlight the dominant influence of monsoon precipitation, baseflow, and glacier melt in maintaining streamflow and provide valuable insights for future water resource planning, climate change impact assessment, and disaster risk management in data-scarce mountainous regions.

3.4.2.11 Russia

Nasonova et al. (2018) projected future streamflow changes for the Northern Dvina, Kolyma, and Indigirka rivers through 2100 using the SWAP hydrological model, driven by climate data from the Russian Academy of Sciences. Under RCP4.5 and RCP8.5 scenarios, mean annual runoff is expected to increase by 16–23%, 16–28%, and 12–26%, respectively, compared to historical baselines. These projections suggest significant increases in freshwater availability, especially under high-emission scenarios.

3.4.2.12 Turkey

Yildirim et al. (2021) used the HYPE model to assess climate change impacts on the Alata River Basin under RCP8.5. After calibration, simulations for 2021–2040, 2046–2065, and 2081–2100 showed rising temperatures (up to 4.11°C) and declining precipitation. Peak discharge shifted from March to February in early periods. Initial increases in runoff due to snowmelt were later followed by reductions caused by increased evapotranspiration and decreased snow accumulation.

3.4.2.13 Qatar

Baalousha (2016) estimated groundwater recharge using a water balance model and Monte Carlo simulations. By applying Darcy's Law, annual recharge was calculated at about 58.7 million m³. These findings support aquifer storage and recovery (ASR) planning in the country.

3.4.2.14 Saudi Arabia

Masoud et al. (2024) estimated groundwater recharge in Jazan Province using satellite imagery, the SWAT model, and GIS. Recharge ranged from 0.002 to 85 mm km^{-2} year^{-1}, with an average of 2.5 mm km^{-2} year^{-1}.

3.4.2.15 Kuwait

Al-Huwaishel et al. (2022) evaluated the feasibility of ASR using treated wastewater in the Dammam aquifer. Using a MODFLOW–MT3DMS model, the scenario with 650 m³ day^{-1} injection and 1000 m³ day^{-1} recovery showed the best efficiency and water quality. ASR was found to be a promising solution for regional water scarcity.

3.4.2.16 South Korea

Ware et al. (2023) applied a coupled SWAT-MODFLOW–TWTFM model in the Anyang watershed. The model improved streamflow simulations and showed that most groundwater recharge occurs during the wet season. Surface and lateral flows were also major contributors to the water balance.

3.4.2.17 Thailand

Ditthakit et al. (2021) demonstrated that machine learning methods, MLR, Random Forest, and M5 Model Tree, can effectively estimate monthly runoff in ungauged basins in the southern part of the country.

3.4.2.18 Taiwan

Ngo et al. (2024) used the SWAT-MODFLOW model to assess streamflow and groundwater recharge in the Choushui River Alluvial Fan. Recharge may increase by up to 66% under RCP2.6 (2061–2080), but later periods show decreases of 23–42% under all scenarios, reflecting diverse climate change impacts.

3.4.3 CASE STUDIES FROM AMERICA

In the Americas, where climate change, population growth, and land-use changes continue to intensify pressure on water resources, hydrological models offer valuable insights into potential impacts on water availability. By simulating future hydrological conditions under various scenarios, these models support stakeholders in making informed decisions about water management, irrigation strategies, and the sustainable use of water resources in agricultural regions. In recent years, several case studies across the Americas (Table 3.6) have demonstrated the usefulness of hydrological

---

TABLE 3.6

Overview of Case Studies Assessing the Impacts of Climate Change on Agricultural Water Resources in Selected American Countries

Country/Region	Study Area	Hydrological Model(s)	References
North America	Great Lakes Basin	SWAT	Myers et al. (2023)
Canada	McKenzie Creek watershed	GSFLOW	Deen et al. (2025)
USA	331 river basins	Budyko method	Wolkeba et al. (2023)
Mexico	Mixteco River Basin	SWAT	Colín-García et al. (2024)
Brazil	Jari River Watershed	SWAT	Rufino et al. (2025)
Argentina	Upper Sali-Dulce watershed	InVEST–AWY	Nunez et al. (2024)
Peru	Quiroz-Chipillico watershed	WEAP	Flores-López et al. (2016)
Colombia	Meta River Basin	InVEST–AWY	Valencia et al. (2024)
Ecuador	Pita River Basin	SWAT	Senent-Aparicio et al. (2024)
Uruguay	Laguna del Sauce catchment	SWAT	Aznarez et al. (2021)
Bolivia	Alto Beni region	-	Fry et al. (2012)
Costa Rica	Five tropical catchments	GR2M	Mendez et al. (2022)
Paraguay	Paraguay River Basin	SWAT+	Barresi Armoa et al. (2024)
El Salvador	Guajoyo River Basin	SWAT	Blanco-Gómez et al. (2019)
Cuba	Caut River Basin	SWAT	Montecelos-Zamora et al. (2018)

models in evaluating water security and guiding adaptive strategies to ensure long-term agricultural sustainability.

3.4.3.1   Northern America

Myers et al. (2023) explored the effects of rain-on-snow (ROS) melt events on snow-pack storage and flood risks in the Great Lakes Basin. Using the SWAT model under RCP4.5, they compared the periods 1960–1999 and 2040–2069. They found a 30% reduction in ROS melt in warmer southern subbasins and less than 5% in colder northern ones. Rainfall increases reduced snowpack, raising the winter/spring rain-to-snow ratio from 1.5 to 1.9. Areas below −2°C were resilient, while those near freezing were more sensitive. ROS melt timing affected spring water yield but had a limited impact in summer.

3.4.3.2   Canada

Deen et al. (2025) assessed BW and green water (GW) scarcity in the McKenzie Creek watershed using the GSFLOW model under RCP4.5 and RCP8.5. BW scarcity could reach moderate levels if maximum withdrawals occur, risking ecosystem degradation. GW scarcity is projected to increase steadily, especially in the western watershed, potentially increasing dependence on BW and stressing downstream availability.

3.4.3.3   United States of America

Wolkeba et al. (2023) used the Budyko method to analyze water availability in 331 river basins under SSP585 and SSP126. They found that 43% of basins under SSP585 and 28% under SSP126 may face increased water scarcity, especially in sub-humid regions. Increased potential evapotranspiration, especially under SSP585, was the main driver, surpassing precipitation changes. The findings call for long-term water management and improved projections.

3.4.3.4   Mexico

Colín-García et al. (2024) evaluated the Mixteco River Basin under SSP245 and SSP585. Projected annual rainfall may drop by 83.71–225.83 mm, with temperatures rising by 2.57–4.77°C. SWAT simulations showed that water yield may decline by 47.40% and 61.01%, respectively, under moderate and high emissions scenarios. The results stress the urgency of measures to reduce climate-related water stress.

3.4.3.5   Brazil

Rufino et al. (2025) applied the SWAT model in the Jari River Watershed to examine climate change and deforestation effects on BW availability (2020–2050). Surface runoff may rise by 18 mm, while groundwater recharge could decrease by 20 mm or increase by up to 120 mm. These shifts could worsen streamflow variability, raise flood risks, and reduce dry-season water availability, impacting ecosystems and human use alike.

3.4.3.6   Argentina

Nunez et al. (2024) used the InVEST–AWY model to study the upper Sali-Dulce watershed. The montane zone, 40% of the area, contributes 80% of the water yield.

Climate change alone could increase yield by 21–75%, while reforestation might reduce it by 15%, and agricultural expansion could raise it by up to 40%. The results highlight the importance of land cover in hydrological responses.

3.4.3.7 Peru

Flores-López et al. (2016) developed a WEAP model for the Quiroz-Chipillico watershed. Although climate change influences water production, land use change is likely the dominant factor shaping hydrological outcomes, with major implications for sustainable water resource management.

3.4.3.8 Colombia

Valencia et al. (2024) applied the InVEST–AWY model with 13 CMIP6 projections in the Meta River Basin. Under SSP2-4.5 and SSP5-8.5, water yield may rise by 24% and 19%, respectively, mainly from increased rainfall in lowlands. Yet, subregional variation exists, slight yield declines in the upper Meta, gains in Casanare, and sharp reductions in the South Cravo subbasin, pointing to possible localized water stress.

3.4.3.9 Ecuador

Senent-Aparicio et al. (2024) used the SWAT model in the Pita River Basin under RCP4.5 and RCP8.5 (2040–2099). Mid-term projections suggest streamflow might increase due to higher rainfall, but temperature-driven evapotranspiration could outweigh this, reducing recharge and streamflow, thus weakening groundwater's future role.

3.4.3.10 Uruguay

Aznarez et al. (2021) applied the SWAT model to the Laguna del Sauce catchment. Under high-emission scenarios, higher temperatures, seasonal changes, and intense rainfall could lower water quality, increase erosion, and destabilize freshwater availability, threatening agricultural water use and crop yields.

3.4.3.11 Bolivia

In the Alto Beni region, Bolivia, Fry et al. (2012) assessed the impacts of climate change and agricultural expansion on spring-fed water supplies. Using satellite precipitation data and hydrological modeling across eleven watersheds, they found that while runoff may fluctuate depending on the scenario, groundwater recharge is projected to decline consistently, by up to nearly 100% by the end of the century. Climate change had a far greater impact than land use change, challenging local perceptions that farming alone is responsible for reduced spring flow. They also found that, although overall groundwater use appears sustainable, limited recharge zones may struggle to meet future water demands, especially for agriculture, without alternative sources.

3.4.3.12 Costa Rica

Mendez et al. (2022) evaluated the impacts of climate change on streamflow in five tropical catchments using the spatially lumped conceptual rainfall-runoff GR2M model. Projections were based on low (RCP 2.6), moderate (RCP 4.5), and high (RCP 8.5) emission scenarios for near (2011–2040), mid (2041–2070), and far (2071–2100) futures. Results showed increased runoff under RCPs 2.6 and 4.5 in the near and mid-terms but decreased runoff under RCP 8.5 by the end of the century. Streamflow changes were mainly driven by shifts in precipitation, indicating increased risks of both floods and droughts.

3.4.3.13 Paraguay

Barresi Armoa et al. (2024) investigated the hydrology of the Paraguay River Basin, part of the La Plata Basin in South America, where streamflow is influenced by high evapotranspiration, gentle slopes, and the Pantanal wetlands. Using the enhanced SWAT+ model with connectivity and Landscape Units, they simulated water exchanges between uplands, floodplains, and channels. Calibrated with 1990–2020 discharge data, the model showed that floodplains store 56.5% of annual flow, with approximately 61% of runoff routed through them, demonstrating improved performance of the SWAT+ model for floodplain-dominated systems.

3.4.3.14 El Salvador

Blanco-Gómez et al. (2019) used the SWAT model to evaluate climate change impacts on water availability and drought in El Salvador's Guajoyo River Basin, a region highly vulnerable to climate change. Under moderate (RCP 4.5) and high (RCP 8.5) emission scenarios, projections showed declining rainfall and rising temperatures by 2050 and 2100. Results indicated reduced water availability and more severe droughts compared to the 1975–2004 baseline. These insights support local water management planning under future climate conditions.

3.4.3.15 Cuba

Montecelos-Zamora et al. (2018) applied the SWAT model for an initial assessment of climate change impacts on water resources in the Cauto River Basin. Calibrated and validated for 2001–2010, the model performed well in two subbasins. Under the high-emission RCP 8.5 scenario, simulations for 2015–2039 projected a 1.5°C temperature rise and a 38% drop in rainfall, resulting in streamflow reductions of up to 61% and aquifer recharge declines of 58%. The findings raise major concerns for regional water security.

3.4.4 CASE STUDIES FROM AFRICA

Water security is an increasing concern across Africa, where agriculture plays a central role in sustaining national economies. Growing climate variability, population pressure, and land-use changes are placing additional strain on already limited water resources (Musse, 2021). In this context, hydrological modeling has emerged as a powerful tool for understanding, managing, and planning agricultural water use. These models enable researchers and decision-makers to simulate complex hydrological processes, evaluate the impacts of climate change, and develop informed strategies for sustainable water management.

The following case studies from various regions of the continent (Table 3.7) illustrate how hydrological modeling has helped enhance water security in agricultural systems, providing evidence-based insights that support policymaking, irrigation planning, and climate adaptation efforts.

TABLE 3.7
Overview of Case Studies Assessing the Impacts of Climate Change on Agricultural Water Resources in Selected African Countries

Country	Study Area	Hydrological Model	References
Ethiopia	Kobo-Golina River catchment	SWAT+	Abate et al. (2024)
South Africa	Eerste River	Pitman	Du Plessis and Kalima (2021)
Morocco	Loukkos Basin	SWAT	Acharki et al. (2023)
Algeria	Cheliff, Tafna, and Macta basins	GR2M	Hadour et al. (2020)
Tunisia	Medjerda River	WEAP	Rajosoa et al. (2022)
Tanzania	Usangu catchment	SWAT	Mollel et al. (2023)
Kenya	Nzoia catchment	SWAT	Githui et al. (2009)

3.4.4.1 Ethiopia

Abate et al. (2024) investigated the future dynamics of blue and GW resources in the Kobo-Golina River catchment, located in the Upper Danakil Basin. Using the SWAT+ model, they evaluated the impacts of climate change on water availability under two socioeconomic pathways (SSP2–4.5 and SSP5–8.5) across three future periods: 2015–2044, 2045–2075, and 2076–2100. Their findings revealed a consistent decline in BW and an increase in GW under both scenarios, with precipitation exerting a more pronounced influence on BW availability. These changes may have significant implications for agricultural water security in the region, as shifts in water availability could affect crop production, irrigation planning, and the long-term sustainability of water resources in this highly climate-sensitive landscape.

3.4.4.2 South Africa

In Stellenbosch, Western Cape, Du Plessis and Kalima (2021) investigated the potential impacts of climate change on water availability in the Eerste River, an essential resource for nearby agricultural areas. Using the Pitman model, the study simulated future river flow under moderate (RCP4.5) and high (RCP8.5) emission scenarios for two future periods: 2022–2057 and 2058–2093. Results indicated a projected decrease in rainfall by 2–8% and an increase in evaporation by 6–15%, leading to a significant reduction in river flow of 8–18%. These climatic shifts are likely to raise irrigation demand by 12–29%. The study recommended expanding the storage capacity of existing farm dams and adopting effective water demand management strategies to build resilience against climate-induced water stress.

3.4.4.3 Morocco

In northwestern Morocco, Acharki et al. (2023) evaluated the impacts of future climate on streamflow in the Loukkos Basin using the SWAT model. Streamflow projections for the 2021–2100 period, under both RCP (4.5 and 8.5) and SSP (SSP2–4.5 and SSP5–8.5) scenarios, were compared to a 1981–2020 baseline. The results indicated a decline in streamflow, with the most significant reduction projected under SSP5–8.5 by the end of the century.

3.4.4.4   Algeria

In northwest Algeria, Hadour et al. (2020) investigated the sensitivity of river flows to future climate change in the Cheliff, Tafna, and Macta basins using the GR2M model under RCP4.5 and RCP8.5 scenarios. An analysis of the 1970–1999 period revealed a notable decline in monthly flows, attributed to reduced precipitation and increased evapotranspiration. Future projections for 2039, 2069, and 2099 indicated continued decreases in winter flows, while spring flows were forecasted to moderately increase under RCP8.5 but decline under RCP4.5.

3.4.4.5   Tunisia

Rajosoa et al. (2022) used the WEAP model to assess the impacts of RCP2.6 and RCP6.0 climate change scenarios on the hydrological regime of the Medjerda River Basin, Tunisia's largest surface water resource. Their findings projected a sharp increase in water demand, from 218 Mm³ in 2020 to 395 Mm³ by 2050, exceeding the available supply, particularly during drought periods. Water shortages were projected to triple relative to baseline levels, posing serious risks to water security in adjacent agricultural zones. Groundwater reserves were also expected to decline significantly, especially between 2045 and 2050.

3.4.4.6   Tanzania

In the Usangu catchment (southwest of the country), an essential rice-growing area, Mollel et al. (2023) used the SWAT model to show that, relative to the 1990–2014 period, projected changes under moderate (SSP2–4.5) and high (SSP5–8.5) emission scenarios over 2030–2060, including a 7–17% increase in annual rainfall, a 0.6–2°C rise in temperature, and up to 30% more evapotranspiration, could reduce water yield and groundwater recharge by 7–26%, emphasizing the significant effect of climate change on the region's hydrological balance.

3.4.4.7   Kenya

In western Kenya, Githui et al. (2009) assessed the potential effects of future climate change on streamflow in the Nzoia catchment, Lake Victoria basin, using the SWAT model. Results showed increased annual rainfall across all climate change scenarios, with monthly variations. Rainfall in the 2050s is expected to be higher than in the 2020s, while temperature increases will be more pronounced, ranging from 0°C to 1.7°C. The change in mean annual rainfall (2.4–23%) is projected to lead to streamflow changes of 6–115%. This could have implications for ecosystem services in the region.

3.4.5   Case Studies from Australia and New Zealand

In Australia, the review by Saha and Zeleke (2015) highlighted that, despite certain limitations, the SWAT model has been applied across a broad range of significant areas, from basic water balance assessments to more complex evaluations of environmental policies and water markets. The findings suggest that the model holds considerable potential for explaining Australia's hydrological dynamics, supporting sustainable water resource management, and enhancing agricultural water security.

By improving understanding of water availability and distribution, the model could play a crucial role in addressing the region's water scarcity challenges and ensuring the long-term sustainability of agricultural practices.

In New Zealand, Collins (2020) employed a climate-hydrology modeling cascade to project changes in river flow for the late 21st century (2080–2099), comparing them to the reference period of 1986–2005. Covering over 43,000 river locations, their study identified significant increases in mean annual flows, surpassing 20% in some areas, along the west and south of the South Island. In contrast, the north and east of the North Island were projected to experience decreases exceeding 10%, with the most pronounced changes occurring under higher emission scenarios. This extensive study provides valuable insights into the potential long-term impacts of climate change on water availability across New Zealand.

3.5   LIMITATIONS

Despite ongoing advancements, hydrological models continue to face some uncertainties. These uncertainties are largely due to incomplete input data, calibration challenges, and simplified model structures. Often, gaps in data, complex algorithms, and the need to approximate processes across different scales contribute to these issues. As such simplifications are unavoidable, and uncertainty remains a fundamental aspect of modeling. Tackling this requires focused uncertainty analysis (UA), which supports model refinement, guides future data collection efforts, and helps assess the reliability of predictions (Moges et al., 2021).

Advancing hydrological modeling for sustainable agricultural water management increasingly relies on the integration of diverse modeling approaches and emerging technologies. The combination of different models offers a more comprehensive method to capture the complexity of real-world hydrological systems. Meanwhile, integrating artificial intelligence with both mechanistic and empirical models enhances predictive capabilities by enabling data-driven parameterization. Additionally, incorporating remote sensing, geographic information systems (GIS), and global positioning systems (GPS) improves spatial resolution and facilitates the efficient management of large, distributed datasets (Liangliang & Daoliang, 2014). Collectively, these advancements strengthen scenario-based assessments and promote more informed, adaptive, and resilient water management strategies, particularly crucial in agriculture, the world's largest water-consuming sector.

3.6   CONCLUSION

This chapter has demonstrated the critical role of hydrological modeling in assessing agricultural water security under climate change. Drawing on case studies from various regions, it is evident that hydrological models, particularly SWAT, provide valuable insights into the complex interactions between climate, water resources, and farming practices. These tools enable the simulation of long-term climate impacts, the assessment of both surface and subsurface water dynamics, and the identification of areas vulnerable to water scarcity. Common trends observed across the case studies include increasing water stress, seasonal shifts in water availability, and intensified competition for limited resources. These challenges underscore the urgent need for adaptive, evidence-based water management strategies. Hydrological modeling not only deepens scientific understanding but also supports informed decision-making for policymakers, planners, and local communities. As climate change continues to reshape hydrological patterns, the continued use and advancement of these models will be essential for achieving sustainable and resilient agricultural systems worldwide. Future research should focus on improving model integration, incorporating emerging data sources, and enhancing scenario-based assessments to better support sustainable water management in agriculture.

4 Barriers to Agricultural Water Security
A Global Analysis of Climatic Risks and Non-Climatic Constraints

4.1 INTRODUCTION

Achieving water security in agriculture is of paramount importance for several reasons. Foremost, an adequate water supply, whether through rainfall or irrigation, is crucial for optimal crop growth and yields, thereby significantly contributing to global food security (Young et al., 2021). Additionally, a reliable water supply is essential for sustainable livestock production, another key contributor to food security (von Keyserlingk et al., 2016). Furthermore, agricultural water management extends beyond the goal of water security; it also seeks to maintain the long-term productivity of farming systems while promoting environmental protection. This includes safeguarding water resources, enhancing soil health, and preserving biodiversity (Shit et al., 2024). Moreover, adequate access to agricultural water can foster socioeconomic growth and contribute to poverty reduction, particularly in rural communities, such as those in Sub-Saharan Africa (Mupaso et al., 2023).

Achieving long-term agricultural water security depends not only on the availability of water but also on a range of climatic (e.g., Kharraz et al., 2012) and non-climatic (e.g., Morton, 2015) factors. While climatic uncertainties and extreme weather events, such as droughts, floods, windstorms, and heatwaves, pose significant challenges to farming systems, non-climatic barriers, including unsustainable water use, governance failures, inadequate infrastructure, socioeconomic inequalities, and policy misalignments, often exacerbate the situation. The interplay between these climatic and non-climatic factors, combined with the rising demand for water over recent decades, driven by population growth, urbanization, increased consumption, agricultural intensification, and land use changes, has led to immense pressure on water resources and undermined their long-term security. This has contributed to a severe water crisis in many regions, particularly in arid and semi-arid areas (de Jager et al., 2022; Novoa et al., 2024). This, in turn, may hinder progress toward achieving the United Nations Sustainable Development Goals (SDGs), such as Zero Hunger (SDG 2), Clean Water and Sanitation (SDG 6), Good Health (SDG 3), and Life on Land (SDG 15) by 2030 (Taka et al., 2021).

DOI: 10.1201/9781003660521-4

Understanding the level of agricultural water security is essential for developing effective resilience and resource management strategies. In this context, the chapter aims to provide a comprehensive examination of the key challenges affecting water availability and use in farming systems, with a focus on both climatic and non-climatic drivers. Based on case studies from Europe, Africa, Asia, the Americas, and Australia, it examines how these drivers can influence agricultural water management. Specifically, this chapter pursues the following objectives:

- To identify the primary climatic and non-climatic factors affecting agricultural water security.
- To illustrate, through international case studies, how these factors may lead to water stress.

4.2   CLIMATIC RISKS TO AGRICULTURAL WATER SECURITY

Climate poses a wide range of risks to the agricultural sector, with water resources being among the most critically affected due to their high sensitivity to weather variables. From extreme events such as floods, droughts, and heatwaves to gradual shifts like rising temperatures, rainfall variability, and increased evapotranspiration, these climate-related stressors disrupt the availability, distribution, and quality of water. This section explores how these climate risks are reshaping water dynamics in agriculture and what this means for the future of farming systems.

4.2.1   IMPACTS OF DROUGHTS AND HEATWAVES

Extreme droughts and heatwaves are becoming more frequent and intense globally due to global warming, putting increasing pressure on freshwater resources (e.g., Payus et al., 2020). Perkins-Kirkpatrick and Gibson (2017) found that for every degree of global warming, the duration of heatwaves in the mid to high latitudes of the Northern Hemisphere increases by about 10 to 15 days, with their intensity rising by approximately 1–1.5°C. On a global scale, compared to the year 2000, the frequency and duration of droughts have risen by roughly 30% (UNDRR, 2021). Sub-Saharan Africa, Australia, and parts of the Americas regularly experience severe drought and heatwave events, highlighting the urgent need for effective water resource management under climate change (WMO, 2020). Furthermore, countries like Ireland, where droughts were once considered uncommon, are now facing them with increasing regularity (Augustenborg et al., 2022).

These hydroclimatic extremes disrupt the water cycle by accelerating soil moisture loss and surface water evaporation, leading to reduced water volumes in surface water bodies such as rivers, streams, and lakes (Payus et al., 2020). As a result, water scarcity intensifies, putting greater pressure on freshwater resources. The implications for water management are multifaceted (Figure 4.1), as these events can lead to: (1) Reduced water availability (i.e., water scarcity), (2) increased agricultural water demand, and (3) deterioration of water quality to levels unsuitable for use (van Vliet et al., 2024).

```
┌──────────────────────────────────────────────────────────────┐
│             Implications of droughts and heatwaves             │
└──────────────────────────────────────────────────────────────┘
        ⇓                        ⇓                        ⇓
┌──────────────────┐    ┌──────────────────┐    ┌──────────────────┐
│  Reduced water   │    │ Increased water  │    │  Water quality   │
│   availability   │    │     demand       │    │  deterioration   │
└──────────────────┘    └──────────────────┘    └──────────────────┘
```

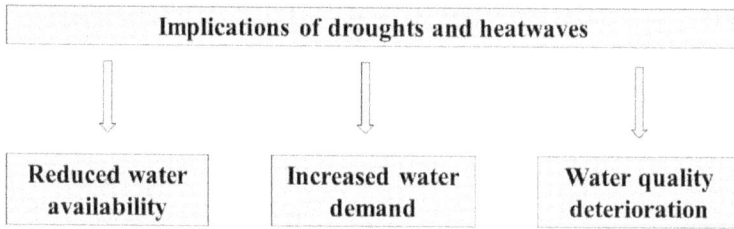

FIGURE 4.1   Implications of droughts and heatwaves on agricultural water resources.

4.2.1.1 Water Scarcity

Droughts are extended periods of insufficient water availability, whether from the atmosphere, surface sources, soil, or groundwater, that result in water scarcity and disrupt human activities (IDMP, 2022). In the agricultural context, drought refers to the lack of adequate moisture, irrigation water, or rainfall required to sustain normal farming operations. It typically occurs when water demand surpasses supply, leading to agricultural water scarcity.

Droughts are generally classified into three main types: (1) Meteorological drought, which arises when precipitation remains below the long-term average over a given period; (2) Agricultural drought, characterized by insufficient moisture in the root zone of crops, thereby affecting plant development and yield; and (3) Hydrological drought, defined by reduced water levels in rivers, lakes, reservoirs, and groundwater systems (Nairizi, 2017).

To monitor these drought types over a specific region and timeframe, researchers commonly recommend indices such as the Standardized Precipitation Index (SPI) and the Standardized Precipitation Evapotranspiration Index (SPEI) (e.g., Burić et al., 2024). SPEI offers an advantage over SPI by incorporating potential evapotranspiration (PET) alongside precipitation, thus providing a more comprehensive reflection of drought conditions (Filipović & Tošić, 2024; Burić et al., 2024).

In rainfed agriculture, the primary impact of drought is soil water deficit, where insufficient green water (i.e., soil moisture) hampers healthy crop development. In irrigated systems, drought exacerbates blue water scarcity by reducing the availability of water for irrigation. Notably, Rosa et al. (2020) reported that 76% of global croplands experience green water scarcity, while 68% of irrigated croplands face blue water scarcity, each persisting for at least one month per year. Furthermore, Liu et al. (2022b) projected that, relative to the 1981–2005 baseline, approximately 80% of global croplands will face agricultural water security challenges by 2050 because of water scarcity and increased crop water requirements (CWR). This alarming projection highlights the urgent need for proactive measures.

Water scarcity in croplands limits water availability for crops, ultimately reducing crop growth and yield. Kim et al. (2019) found that, due to drought events between 1983 and 2009, approximately 75% of global agricultural areas experienced crop yield losses, resulting in an estimated economic loss of around 166 billion US dollars. Yield losses are exacerbated when heatwaves and droughts occur simultaneously (Sutanto et al., 2024). During drought, plants close their stomata to conserve water, which also limits carbon dioxide intake and reduces photosynthesis (Li et al., 2020). Prolonged water stress can further impair the photosynthetic electron transport chain, leading to cellular damage and long-term down-regulation of photosynthesis (Li et al., 2020). When combined with heatwaves, the stress intensifies, as high temperatures increase evapotranspiration and disrupt cellular structures, exacerbating oxidative stress and further reducing plant productivity. These compound effects are more damaging than individual stressors, resulting in more severe crop yield losses (Sutanto et al., 2024).

4.2.1.2   Increased Water Demand

Drought often leads to severe stress on CWR. Rising drought frequency, coupled with anticipated increases in PET, is expected to significantly alter the length of the crop growing season (LGP) and increase crop water demands (Ray et al., 2018; Abraham & Muluneh, 2022). In rainfed agriculture, reduced rainfall during droughts diminishes soil moisture availability, causing crops to experience water stress earlier in the growing season. To mitigate the negative effects of this stress and balance CWR, supplemental irrigation is often applied (Gebremedhin et al., 2023), leading to increased water supply needs. In irrigated systems, the absence of rainfall increases dependence on irrigation, while rising temperatures further accelerate evapotranspiration, necessitating higher amounts and more frequent irrigation to meet CWR. Consequently, drought conditions raise water demand in both rainfed and irrigated agriculture, posing a significant challenge for sustainable water resource management under ongoing and future climate change.

4.2.1.3   Water Quality Deterioration

Long-term water quality monitoring is crucial for assessing the impact of drought on freshwater resources (Burt et al., 2014; Mosley, 2015). These datasets have supported a broad range of studies, revealing how hydrological droughts can compromise freshwater resources. Mosley (2015), in particular, offers an in-depth review of these processes. Reduced water availability during drought often results in higher salinity due to limited dilution and increased solute concentration. Increased air temperatures and longer water retention periods can enhance thermal layering, supporting algal proliferation, including harmful cyanobacteria, and contributing to oxygen depletion. In lakes, decreased flushing combined with heightened biological activity typically leads to higher nutrient loads, turbidity, and algal biomass. Conversely, rivers and streams with minimal anthropogenic inputs may show reduced nutrient and turbidity levels, largely due to the interruption of external sources and a stronger influence of internal mechanisms such as nutrient uptake, sedimentation, and denitrification. On the other hand, in systems impacted by point-source pollutants, limited water flow during droughts often leads to more pronounced declines in water quality, primarily due to the reduced capacity for nutrient dilution (Mosley, 2015).

4.2.2   Impacts of Floods

Floods can significantly influence water security in agricultural areas, with both positive and negative implications for water availability and quality (Figure 4.2). In

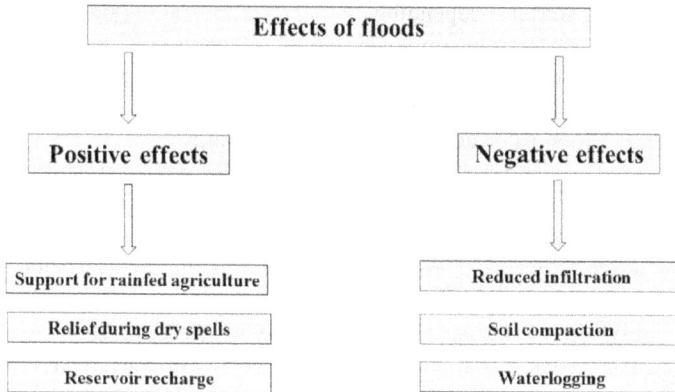

FIGURE 4.2   Impacts of floods on agricultural water resources.

certain contexts, particularly in arid and semi-arid regions, flooding may temporarily enhance water security by recharging reservoirs, supporting rain-fed cropping systems, and alleviating the effects of dry spells. These conditions can also benefit the following planting season by increasing soil moisture and depositing nutrient-rich sediments that support plant growth (FAO, 2011).

However, in many other cases, excessive rainfall and prolonged standing water on agricultural land can result in waterlogging, reduced infiltration, soil compaction, and disruption of the hydrological balance within the root zone. Such conditions can negatively affect soil health and reduce crop productivity (FAO, 2011).

Although short-term benefits exist, the long-term consequences of floods on water security are often more detrimental, particularly regarding water quality and the resilience of water infrastructure (Barbetta et al., 2022). Floodwaters frequently carry a mixture of pollutants, including sediments and agrochemicals, which can degrade surface and groundwater quality. This degradation poses serious risks to farming operations. Hrdinka et al. (2012) argue that even brief flood events can lead to more severe water quality deterioration than prolonged droughts.

Crucial infrastructure systems supporting agricultural water supply, such as reservoirs, wells, and irrigation networks, are also vulnerable to damage during flood events (Barbetta et al., 2022). Elevated water levels can compromise dam integrity, increase turbidity, and reduce reservoir storage through sedimentation. Flooded well fields may suffer from equipment failure and aquifer contamination by polluted floodwater, thereby jeopardizing groundwater safety. Moreover, water monitoring systems may be disrupted, further complicating efforts to ensure consistent water quality and supply (Barbetta et al., 2022).

4.2.3   IMPACTS OF INCREASING TEMPERATURES

Rising temperatures, particularly in arid and semi-arid agricultural regions, are associated with increased water evaporation, soil moisture depletion, and elevated crop evapotranspiration rates (OECD, 2014). These climatic shifts can either extend or shorten growing seasons, depending on the geographical region. In high-latitude areas, longer growing seasons are becoming more common (Gornall et al., 2010), whereas in other regions, the trend may lead to shorter growing periods, potentially reducing CWR, lowering irrigation demands, and supporting more efficient use of water resources (Soomro et al., 2023). In predominantly rainfed agricultural systems, an extended growing season could enhance yields, provided that water remains sufficiently available throughout the period (Mueller et al., 2015).

In northern regions such as Canada, Russia, Finland, Norway, and Alaska, the response of water resources to rising temperatures is particularly notable. These areas have recently experienced unprecedented warming, resulting in accelerated snowmelt and significant glacier retreat (e.g., Nury et al., 2022). Between 1980 and 2015, it is estimated that global glaciers experienced an average annual area shrinkage of 0.18% and a mass loss rate of 0.25 m water equivalent per year, illustrating the widespread impacts of ongoing global warming (Li et al., 2019).

The intensified melting of snowpacks and glaciers has led to increased river flows in recent decades (e.g., Lutz et al., 2014), temporarily improving water availability for downstream agricultural use and acting as a buffer against drought (e.g., Pritchard, 2019). However, as global warming progresses and glacier mass continues to decline, the long-term ability of these frozen reservoirs to store and gradually release water is expected to diminish. This poses a serious threat to water availability for downstream communities and agricultural systems (e.g., Liu et al., 2024), highlighting the critical need for forward-looking planning to maintain both agricultural productivity and ecological stability in these vulnerable regions.

4.2.4 IMPACTS OF ERRATIC RAINFALL PATTERNS

Erratic rainfall, characterized by unpredictable timing, intensity, and spatial distribution, can profoundly affect water resources, complicating both water and crop management in agricultural regions. Understanding rainfall patterns is essential for identifying areas experiencing either water surpluses or shortages, thereby enabling more effective planning for agricultural water distribution (Garg et al., 2022).

Over recent decades, the acceleration of the global water cycle, driven by climate change, has led to substantial alterations in precipitation patterns at both seasonal and annual scales. These shifts are reflected in the changing geographic distribution, duration, and intensity of rainfall events, with significant implications for both blue and green water availability (Ehtasham et al., 2024). In regions characterized by unimodal rainfall regimes, increased seasonal variability has disrupted the regular supply of atmospheric moisture, resulting in prolonged dry periods. In contrast, areas that typically receive abundant rainfall are now experiencing shorter yet more intense precipitation events, heightening the risk of flooding and presenting new challenges for agricultural water management and storage infrastructure (Konapala et al., 2020).

Nonetheless, it is important to note that in such high-rainfall regions, intensified rainfall can lead to greater annual runoff, potentially enhancing blue water resources – provided that surplus water is effectively captured and stored. With adequate infrastructure in place, this stored water can be used to support irrigation, reinforce long-term water security, and contribute to improved crop productivity (Ehtasham et al., 2024).

4.2.5   Regional Case Studies

4.2.5.1   Case Studies from Asia

Water resources across Asia are highly vulnerable to climate-related risks, which have already led to rising water stress, scarcity, and supply disruptions, particularly in agricultural regions. Climate change is intensifying these challenges through more frequent and severe floods and droughts. Drawing on various analytical approaches, several studies have assessed the impacts of these climatic threats on agricultural water security. Selected examples are summarized in Table 4.1.

4.2.5.1.1   China

Liu et al. (2023) applied the Agricultural Water Scarcity Index (AWSI), an index incorporating both blue water and soil moisture, to evaluate the impact of changing precipitation patterns on water security. Their study compared water scarcity levels between the 1971–2010 period and a projected future period (2031–2070). The results showed that altered precipitation patterns significantly affect agricultural water availability. In extremely dry years, water scarcity levels could rise by up to 31% relative to the average. For the 2031–2070 period, climate change is expected to exacerbate water shortages, particularly in regions like Inner Mongolia, where scarcity could increase by as much as 200%. The study also identified improved irrigation efficiency as a key adaptation strategy, with the potential to reduce water scarcity by up to 30%.

4.2.5.1.2   Iran

In this semi-arid country, Ashraf et al. (2021) reported that prolonged droughts, especially in the northwestern and central semi-arid to hyper-arid areas, alongside human activities such as aquifer overexploitation, have intensified groundwater depletion. These factors have significantly undermined water security, particularly in agriculture, and have led to widespread socioeconomic impacts. Between 2002 and 2015, Iran experienced an estimated total groundwater loss of around 74 km^3,

TABLE 4.1

Impacts of Climatic Factors on Water Resources: Case Studies from Asian Countries

Country	Climatic Factor	Main Impact	Case Study
China	Change in rainfall patterns	Higher water scarcity	Liu et al. (2023)
Iran	Drought	Groundwater depletion	Ashraf et al. (2021)
Malaysia	Flooding	Water quality deterioration	Ching et al. (2015)
Kazakhstan	Precipitation reduction	Water shortage	Kakabayev et al. (2023)
UAE	Low rainfall	Groundwater depletion	Murad (2010)

with droughts identified as a major contributor. The study issued a stark warning that continued depletion could lead to critical water insecurity across all user sectors and trigger severe socioeconomic instability, highlighting the urgent need for sustainable groundwater governance.

4.2.5.1.3 Malaysia

Ching et al. (2015) employed the Water Quality Index (WQI), which considers variables such as dissolved oxygen, pH, and total suspended solids, to investigate the impact of flood events on the Muar River's water quality. Their findings revealed that the severe floods of January 2007 led to increased pollutant loads from surface runoff, which in turn caused a significant decline in water quality, with the WQI dropping to Class III (polluted). These results underline the importance of implementing effective surface water conservation measures to mitigate the impacts of flooding.

4.2.5.1.4 Kazakhstan

In Kazakhstan, where climate patterns are influenced by the Pacific, Atlantic, Indian, and Arctic Oceans, Kakabayev et al. (2023) observed a notable reduction in annual precipitation across large areas of the country, by 20–40% below the long-term average. This decline signals a growing threat of water scarcity linked to climate change, raising serious concerns about the long-term viability of the agricultural sector, which remains heavily dependent on stable and sufficient water supplies.

4.2.5.1.5 United Arab Emirates

In the UAE, where approximately 70% of groundwater is used for agriculture, prolonged periods of low rainfall, resulting from both the arid climate and the influence of climate change, have reduced groundwater recharge (Murad, 2010). This, coupled with excessive groundwater extraction, has led to a marked decline in groundwater availability, posing serious challenges for agricultural sustainability in the country.

4.2.5.2 Case Studies from Europe

Studies (e.g., Zapata-Sierra et al., 2022) have shown that the resilience of water resources across much of Europe is increasingly at risk due to various climatic pressures, including changing precipitation patterns, increased drought frequency, and more frequent flooding events. These shifts have contributed to declining green water availability, rising water shortages, and worsening water pollution, especially in agricultural regions. Several case studies from across Europe are summarized in Table 4.2.

4.2.5.2.1 Maltese Islands

A significant warming trend of 1.5°C was observed between 1952 and 2022, along with a 1.81% decrease in annual rainfall when comparing 1991–2020 to 1961–1990 (Galdies, 2022; Mifsud Scicluna & Galdies, 2025). These climatic changes have already led to reduced water availability. Projections for 2050 and 2070 suggest further increases in temperature and declines in precipitation, which will intensify aridity and heat stress. These trends are expected to reduce soil moisture even further and place mounting pressure on agricultural systems. Among the regions studied,

TABLE 4.2
Impacts of Climatic Factors on Water Resources: Case Studies from European Countries

Country	Climatic Factor	Main Impact	Case Study
Malta	Warming, less precipitation	Soil moisture loss	Mifsud Scicluna and Galdies (2025)
Romania	Drought	Soil moisture reduction	Mateescu et al. (2013)
Czech Republic	Drought	Soil moisture reduction	Trnka et al. (2015)
Greece	Droughts, floods	Risk of water shortage	Kourgialas (2021)
France	Precipitation intensity	Groundwater recharge decline	Sobaga et al. (2024)

the northern and southeastern parts of the islands are expected to face the highest heat-related stress, posing significant risks to agricultural water security (Mifsud Scicluna & Galdies, 2025).

4.2.5.2.2   Romania
Mateescu et al. (2013) reported that drought conditions affected approximately 7.1 million hectares, or 48% of Romania's total agricultural area. The southern, southeastern, and eastern regions are especially vulnerable, often experiencing extreme and severe soil drought, with water availability falling below 600 m³ ha⁻¹. During extended droughts, average crop yields dropped by 35–60% of their potential, clearly demonstrating the major impact of reduced soil moisture on agricultural productivity.

4.2.5.2.3   Czech Republic
Recent short-term droughts have begun to dry out agricultural soils across the Czech Republic. A 50-year soil moisture analysis (1961–2012) by Trnka et al. (2015) revealed a clear decline in soil moisture, particularly during the spring growing season (May–June), and a 50% increase in drought probability. These findings signal a shift toward drier conditions during critical periods for crops, with significant implications for agricultural output.

4.2.5.2.4   Greece
In northwestern Crete (Greece), a Mediterranean region heavily reliant on agriculture and known for its tree crops, Kourgialas (2021) used the SPI to analyze hydrological extremes between 1960 and 2019. The results showed that approximately 24% of the tree crop area is vulnerable to either drought or flood events. These risks threaten both agricultural water security and food production, highlighting the urgent need for adaptation strategies to reduce the impact of climatic extremes on local agriculture.

4.2.5.2.5   France
In many European countries, groundwater recharge patterns now show spatiotemporal variability influenced by climate, farming practices, and land use. In northeastern

France, Sobaga et al. (2024) analyzed groundwater recharge trends from 1975 to 2020 and found a declining trend, averaging −2.5 mm per year. Importantly, the study revealed that intense rainfall does not necessarily result in effective recharge, as 70% of high-intensity precipitation events led to no recharge at all. Seasonal and land cover differences strongly affected recharge efficiency. For instance, during summer, even under bare soil, only 30% of intense rain events produced significant recharge. In contrast, although winter had fewer intense events, just 8% of the total, they were much more effective in recharging groundwater.

4.2.5.3   Case Studies from the Americas

Studies (e.g., Thomas et al., 2016; Rivera et al., 2021) have shown that water resources across various regions of the Americas are increasingly threatened by recent climatic shifts, including declining precipitation and rising temperatures. These climatic changes have led to reduced groundwater recharge, water quality deterioration, and diminished streamflow, significantly undermining agricultural water security. Below are selected case studies from across the continent (Table 4.3).

4.2.5.3.1   USA

In arid and semi-arid regions, groundwater recharge is highly dependent on the intensity and frequency of precipitation, soil texture, and groundwater depth (Thomas et al., 2016). For example, in the southwestern USA, Thomas et al. (2016) observed that reduced precipitation intensity was associated with a lower recharge-to-precipitation ratio. The findings suggest that when rainfall events are less intense, insufficient infiltration occurs to overcome the soil moisture deficit and reach the water table. Consequently, reduced recharge poses a direct threat to agricultural water security by diminishing both the quantity and reliability of water available for irrigation.

4.2.5.3.2   Canada

In cold and typically water-abundant regions, recent climatic extremes have also affected water resources. Blagrave et al. (2022) investigated the nearshore area of Lake Ontario, Canada, between 2000 and 2018 and found that storms and heatwaves significantly deteriorated water quality. Antecedent heatwaves accounted for

TABLE 4.3

Impacts of Climatic Factors on Water Resources: Case Studies from American Countries

Country	Climatic Factor	Main Impact	Case Study
USA	Precipitation intensity	Groundwater recharge decrease	Thomas et al. (2016)
Canada	Heatwaves, storms	Water quality deterioration	Blagrave et al. (2022)
Argentina	Declining snowpack	Reduced streamflow	Rivera et al. (2021)
Brazil	Declining rainfall	Risk of water shortage	Chagas et al. (2022)
Mexico	Hydrological drought	Risk of water shortage	Vásquez et al. (2022)

approximately 87% of elevated chlorophyll levels, while precipitation and storm events explained around 35% of turbidity extremes. This degradation in water quality may jeopardize agricultural water security by reducing the availability of suitable water for irrigation in adjacent farming regions.

4.2.5.3.3 Argentina

In Central-Western Argentina, where agriculture, particularly viticulture, depends on snow-fed rivers from the Andes, Rivera et al. (2021) reported a marked decline in snowpack between 2010 and 2020 due to reduced winter precipitation. This resulted in a prolonged hydrological drought, accelerating glacier retreat, earlier snowmelt, and reduced streamflow. As lakes and reservoirs shrank, pressure increased on already stressed groundwater reserves. These changes underscore the urgency of improved snow drought monitoring and integrated water management strategies to sustain agricultural productivity.

4.2.5.3.4 Brazil

Chagas et al. (2022) conducted a hydrological analysis using daily streamflow data from 1980 to 2015, revealing that approximately 42% of Brazil, home to some of the world's largest river basins, experienced a drying trend. This was characterized by a decrease in flood events and an increase in drought frequency. The shift, driven largely by declining rainfall and compounded by intensifying agricultural water demand, reflects a disrupted water cycle and growing strain on water resources.

4.2.5.3.5 Mexico

In Mexico, Vásquez et al. (2022) examined the Sonora River Basin and found that hydrological droughts became more severe from 1974 to 2013, particularly toward the end of this period. These droughts, driven by changing climatic conditions, led to reduced blue water availability, threatening agricultural water security in the region.

4.2.5.4 Case Studies from Africa

To further illustrate the connection between agricultural water security and climate, the following case studies from selected African countries (Table 4.4) demonstrate how climatic factors, such as rainfall variability, rising temperatures, and extreme

TABLE 4.4

Impacts of Climatic Factors on Water Resources: Case Studies from African Countries

Country	Climatic Factor	Main Impact	Case Study
Zimbabwe	Less rainfall	Surface water shortage	Mazvimavi (2010)
Burkina Faso	Reduced rainfall	Runoff loss, evaporation	Fowe et al. (2015)
Nigeria	Drought, late rain	Water shortage	Adejuwon and Dada (2021)
Cameroon	Rain shifts, warming	Water loss, low income	Fonjong and Zama (2023)
Algeria	Long-term drought	Water shortage	Meddi et al. (2009)

weather events, influence water availability, demand, and management in agricultural systems.

4.2.5.4.1 Zimbabwe

In Zimbabwe, agricultural practices are heavily reliant on surface water, particularly from large dams, due to limited groundwater resources (Mwadzingeni et al., 2022). A study by Mazvimavi (2010) reported a 10% decline in rainfall over the past century, reflecting a downward trend in dam water levels with serious implications for agricultural water security and food production. Projections indicate that this decline in rainfall will continue, accompanied by rising temperatures, both contributing to reduced water availability for irrigation. This water stress is expected to intensify, as irrigation demands increase due to higher evapotranspiration rates (Mwadzingeni et al., 2022).

4.2.5.4.2 Burkina Faso

In West Africa, where water availability is a limiting factor for agricultural development, small reservoirs are essential for supporting irrigation, livestock, and fish production (Owusu et al., 2022). However, rising temperatures and declining rainfall in recent years have led to significant water losses due to increased evaporation and reduced runoff. A study by Fowe et al. (2015) on the Boura reservoir in Centre-West Burkina Faso showed that a 32% reduction in annual rainfall caused a 50% drop in runoff, which, in turn, reduced water retention. The study also found that approximately 60% of the stored water was lost through evaporation. These findings highlight the urgent need for long-term programmes to help local communities better manage reservoirs and watersheds.

4.2.5.4.3 Nigeria

In Nigeria, the frequent occurrence of prolonged droughts (lasting more than one month) and delayed rainfall in recent decades (Adejuwon & Dada, 2021) has caused significant water shortages. As a result, water resources are often insufficient to meet CWR, particularly during the maturity stage, leading to decreased food security (Abulude & Wahlen, 2024).

4.2.5.4.4 Cameroon

A study by Fonjong and Zama (2023) in Cameroon showed that increasing climate variability over the past 21 years, marked by erratic rainfall patterns and rising temperatures, has contributed to the depletion, reduction, or seasonal unreliability of local water sources. These climatic changes have not only reduced freshwater availability but also complicated agricultural management, particularly affecting farmers' income, with a disproportionate impact on women (Fonjong & Zama, 2023).

4.2.5.4.5 Algeria

Algeria has been experiencing a prolonged meteorological drought for many decades, which has exacerbated the national water crisis, especially in the agricultural sector (Meddi et al., 2009). The ongoing drought has placed additional pressure on already scarce water resources, severely limiting irrigation capacity and threatening agricultural productivity.

4.2.5.5   Case Studies from Australia

Australia, like many other regions around the world, is facing increasing water challenges, such as decreasing availability and growing demand. These challenges stem from both climatic factors (e.g., climate change) and non-climatic factors (e.g., urbanization). In response, researchers have increasingly focused on assessing the effects of these drivers on water resources. One of the most comprehensive reviews is that of Wasko et al. (2024), which examined the impact of climatic factors on surface water resources in the region and reported several important findings. For instance, the authors reported that Australia has warmed at approximately 1.4 times the global average rate. This warming has been accompanied by significant changes in rainfall patterns. Annual rainfall has increased in northern regions while declining across much of the south, particularly along the southeast and southwest coasts. The review also highlighted a rise in short-duration rainfall events and a decline in longer-duration ones, contributing to more frequent large floods. Changes in soil moisture have followed similar patterns, with wetter conditions in tropical regions and drier soils in the south. Additionally, increasing evaporative demand has further reduced water availability. As a result, less water is flowing into rivers and streams, leading to declining surface water volumes and growing risks to water security across Australia, especially for agriculture, which is heavily dependent on the availability of surface water resources.

4.3   NON-CLIMATIC CONSTRAINTS

In addition to climatic factors, several non-climatic drivers significantly influence agricultural water security, including global population growth, rapid urbanization, economic development, changing consumption patterns, agricultural intensification, and land-use change (Figure 4.3). Understanding and addressing these factors

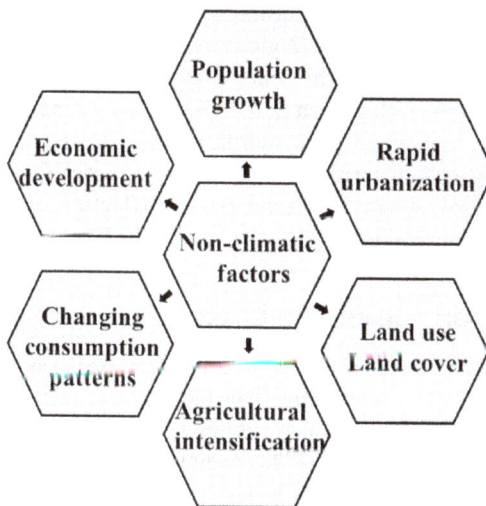

FIGURE 4.3   Non-climatic factors affecting agricultural water security.

is essential for optimizing local farming activities and managing agricultural water resources efficiently.

4.3.1 POPULATION GROWTH

As the global population continues to grow, so does the demand for food and water, which increases the pressure on already limited freshwater resources. According to the United Nations, the global population has increased from 2.5 billion in 1950 to about 8 billion today, intensifying competition among water users, particularly agriculture, which accounts for around 70% of global freshwater withdrawals (Dotaniya et al., 2023). Over the past five decades, farmland has expanded considerably, primarily through the increased use of irrigation, while rain-fed agriculture has remained relatively stable at about 1220 million ha (Chen et al., 2016). This highlights the growing importance of irrigation efficiency. To feed a projected 9.4–10.1 billion people by 2050 (United Nations, 2019), global agricultural output must increase by up to 56% compared to 2010 levels (van Dijk et al., 2021). Achieving this will require highly efficient and sustainable water management, particularly as the effects of climate change exacerbate water stress.

4.3.2 RAPID URBANIZATION

Rapid urbanization is exerting mounting pressure on agricultural water resources by altering land-use patterns, intensifying competition for water, and degrading water quality. Urban growth has often led to a reduction in fertile agricultural land for urban development, which in turn increases household water consumption to meet rising domestic demand, thereby reducing water availability for farming activities (Avazdahandeh & Khalilian, 2021). A study by Avazdahandeh and Khalilian (2021) found that for every 1% increase in urbanization, water use in agriculture decreases by 0.639 mm^3. Furthermore, the expansion of urban areas can weaken the performance of irrigation systems in rural regions, reducing water distribution efficiency and increasing water losses (Rondhi et al., 2024). Runoff from urban surfaces can also carry contaminants such as heavy metals and excess nutrients (e.g., sodium, nitrate, phosphorus), which can pose significant risks to rivers and surface water bodies used for irrigation in surrounding agricultural zones (Tong & Chen, 2002). However, urbanization may also have some positive effects, such as the promotion of urban and peri-urban agriculture (UPA) (Fei et al., 2025).

4.3.3 ECONOMIC DEVELOPMENT AND CHANGING CONSUMPTION PATTERNS

Economic development and changing consumption patterns have significantly impacted agricultural water resources. Over the past five decades, rapid economic growth has driven a shift in dietary preferences toward more water-intensive foods, such as meat (especially beef), dairy, and processed products (Mekonnen & Hoekstra, 2012). This shift has greatly expanded the water footprint of food production, increasing pressure on already limited freshwater supplies and raising the risk of water scarcity, as the food system is the largest global consumer of freshwater (Qin et al., 2019).

Rising incomes and urbanization have also contributed to greater food waste, which indirectly leads to water loss. Coudard et al. (2021) estimate that around 344 million tons of avoidable food waste globally translate into the loss of 82 billion m³ of water. Furthermore, economic development often fuels industrial and infrastructure expansion, intensifying competition for water, especially in water-scarce regions. These trends collectively increase water stress, reduce water availability for farming, and undermine agricultural sustainability. Integrated strategies, such as improving agricultural water conservation, modernizing industrial systems, and encouraging water-saving urban practices, can mitigate these impacts (Jian et al., 2025). A global analysis by Deng et al. (2025) projects that adopting healthier diets, like vegetarian patterns, by 2070 could enhance dietary quality by 30–45%, lower water use by up to 14.7%, and improve food affordability by up to 63%, compared to 2020 levels.

4.3.4   AGRICULTURAL INTENSIFICATION

To meet the demands of a rising global population, accelerated urbanization, ongoing economic growth, and changing consumption patterns, farming systems worldwide are shifting from traditional to more intensive practices (Reardon et al., 2019; Kuchimanchi et al., 2023). While agricultural intensification has enhanced water-use efficiency by producing more food on less land, it has also led to higher water extraction, especially in arid and semi-arid regions that rely heavily on irrigation (Kuchimanchi et al., 2023). For example, a study in Telangana, India, found that agricultural intensification has significantly increased groundwater use, creating a watershed water deficit of about –13.9 Mm³ per year. The dominance of water-intensive crops in these regions has exacerbated groundwater depletion.

In addition to increasing water extraction, intensification often leads to the higher use of agrochemicals and greater reliance on both organic and chemical fertilizers. This may degrade water quality, rendering rivers and other water bodies unsuitable for irrigation (Li et al., 2023). To address these challenges, agricultural intensification must prioritize water conservation, as increased water demand in farming contributes to water scarcity and declining water quality. Without this focus, water scarcity may worsen, water quality will further degrade, and the long-term sustainability of food production systems could be threatened (Benabderrazik et al., 2021).

4.3.5   LAND USE AND LAND COVER

Land use and land cover (LULC) change plays a critical role in shaping both the availability and quality of water, which are essential for agricultural water security. Human interventions, such as deforestation and agricultural expansion, can disrupt natural hydrological processes by altering water flow across the landscape (Shiferaw et al., 2025). These changes affect key components of the water cycle, including evapotranspiration, infiltration, surface runoff, and groundwater recharge. For example, deforestation reduces vegetation interception, modifies soil structure, and exacerbates the effects of droughts, leading to lower water retention and increased runoff (Bagley et al., 2014). This reduces the recharge of groundwater reserves that many farming systems depend on.

For instance, a study by Rufino et al. (2025) in the Jari River Watershed (Brazil) found that deforestation, combined with climate change, could increase surface runoff by more than 18 mm and reduce groundwater recharge by up to 20 mm per month. Such land-use changes undermine the consistent supply of suitable, accessible water needed for sustainable agricultural production, especially in water-stressed regions.

4.4 INSTITUTIONAL AND POLICY BARRIERS

Institutional and policy barriers to water management refer to the challenges within governance systems, laws, regulations, and organizations that hinder the effective planning, allocation, and sustainable use of water resources in agriculture. Below are the most common barriers.

4.4.1 POOR GOVERNANCE

Agricultural water resources are often managed by multiple actors, including decision makers, farmers, formal state institutions, and local water management organizations (Singh et al., 2024). This multi-actor system frequently results in a complex and incoherent governance structure that hampers effective policy formulation, leads to duplication of efforts, and contributes to inefficient water management (Singh et al., 2024). For example, a study by Homobono et al. (2022) conducted in southern Portugal highlighted that the absence of integrated development strategies and poor communication among stakeholders were key barriers to sustainable agricultural water governance. It also highlighted that limited collaboration, particularly between researchers and practitioners, further undermined effective water management. The study concluded that place-specific policies promoting communication and collective action among all involved parties are essential for achieving sustainable agricultural water use.

4.4.2 INADEQUATE POLICIES

In some countries, existing water laws are insufficient to address pressing challenges such as climate change, population growth, rising agricultural demand, and sustainability goals. In regions including Sub-Saharan Africa, Latin America, Southeast Asia, and the Middle East, water policies often remain at the discussion stage, with limited progress in implementation (Cosgrove & Loucks, 2015; Adom & Simatele, 2024). Furthermore, the lack of stakeholder and local community engagement during policy development frequently results in ineffective water resource management (Teweldebrihan & Dinka, 2025). Even where sound policies are in place, corruption can significantly hinder their execution. For example, in Pakistan, a study by Jacoby et al. (2021) revealed that corruption contributed to the failure of decentralizing water management, from bureaucratic agencies to local farmer organizations, as the policy did not achieve equitable water distribution or prevent resource misuse.

4.4.3 INADEQUATE CAPACITY AND MONITORING

Many countries, especially developing ones, lack the necessary tools, technical expertise, staffing, and financial resources to effectively plan, implement, and monitor water management. In addition, the widespread adoption of advanced technologies, such as Artificial Intelligence (AI) and the Internet of Things (IoT), that support efficient water management in agriculture is often hindered by high costs, limited expertise, data challenges, policy and legal constraints, and inadequate infrastructure, especially in developing countries where these limitations are most acute (Parra-López et al., 2025).

4.4.4 INADEQUATE GOVERNANCE OF SHARED WATER RESOURCES

Globally, approximately 260 transboundary rivers (e.g., the Nile, Indus, and Mekong basins) and 450 transboundary aquifers (e.g., the Nubian Sandstone Aquifer System, the Guarani Aquifer, and the North Western Sahara Aquifer System) exist, collectively covering around 50% of the Earth's surface (Sadoff & Grey, 2005; Eckstein & Sindico, 2014; Hossen et al., 2023). These shared water resources are often sources of tension or mismanagement, with inadequate cooperation between riparian states. Such disputes can undermine long-term agricultural water planning, ultimately contributing to water insecurity and regional instability. For example, Hossen et al. (2023) reported that roughly 35% of the world's transboundary river basins lack effective conflict resolution mechanisms, highlighting the urgent need for improved governance and cooperative frameworks.

4.5 CONCLUSIONS

Achieving water security in agriculture is increasingly complex, as water managers must navigate a growing array of climatic and non-climatic challenges. This chapter's findings emphasize that changing weather patterns, such as prolonged droughts, erratic rainfall, and rising temperatures, are significantly complicating the management of agricultural water resources. Simultaneously, non-climatic factors, including unsustainable land use, rapid population growth, agricultural intensification, and socioeconomic pressures, are further straining already limited water supplies. These interconnected challenges not only exacerbate water scarcity but also threaten the reliability and quality of water essential for sustainable food production.

To address these issues, agricultural water managers must continually adapt to and manage a combination of technical, environmental, and structural pressures. This is particularly crucial in vulnerable regions, such as Africa, where these challenges not only degrade water resources but also contribute to socioeconomic hardships for farmers.

5 Recent Trends in Agricultural Water Security
A Global, Regional, and Local Outlook

5.1 INTRODUCTION

In agriculture, where water use is particularly intensive, water security has become an increasingly pressing concern in recent years due to the growing influence of both climatic and non-climatic pressures. Climatic factors, particularly the ongoing impacts of climate change, are affecting both blue and green water resources through altered precipitation patterns, melting glaciers, and more frequent hydrological droughts (Abbas et al., 2022). These climate-related shifts, combined with intensifying non-climatic drivers such as accelerated urbanization, changing land use patterns, rising living standards, the expansion of irrigated areas, and rapid population growth (Mishra et al., 2021; Ingrao et al., 2023), are significantly increasing demand for freshwater while simultaneously undermining its availability. As a result, agricultural water withdrawals have risen sharply in many regions worldwide. For instance, in India, agricultural water withdrawals doubled between 1975 and 2010 (Ritchie & Roser, 2018).

Farmers, particularly those dependent on irrigation, are currently facing increasing difficulties in accessing sufficient quantities of freshwater of suitable quality for sustainable agricultural production (Ingrao et al., 2023). Globally, it is estimated that about 60% of irrigated agriculture is subject to extremely high water stress (Kuzma et al., 2023). Even in regions where water scarcity is not widespread, such as parts of Russia and Ukraine, armed conflict and political instability have exacerbated agricultural water insecurity by damaging water infrastructure, degrading water quality, limiting access to irrigation systems, and disrupting institutional water governance (Schillinger et al., 2020).

In light of these escalating pressures, and as a critical first step toward the effective management of increasingly scarce water resources, numerous researchers and international agencies have investigated agricultural water security trends at both global and local scales (e.g., Shiklomanov & Rodda, 2003; Huang et al., 2018; Liu et al., 2022b; Sun et al., 2024). These assessments, typically grounded in historical patterns of water withdrawal and demand, have generated a range of important

 DOI: 10.1201/9781003660521-5

insights, including the identification of seasonal and spatial variations in agricultural water security, the mapping of water-stressed hotspots, and evaluations of the balance between water availability and agricultural needs. These efforts have enhanced monitoring capabilities, enabling better detection and comparison of water stress levels across different regions. Moreover, they have brought agricultural water security to the forefront of global sustainability discussions, particularly in relation to the interconnected challenges of food security, environmental degradation, and socio-economic vulnerability. Building on this, the objective of this chapter is to synthesize the findings of these assessments to better understand the current state of agricultural water security at both global and local levels. This analysis aims to contribute to a more comprehensive understanding of the evolving global-local dynamics that are shaping patterns of agricultural water use, consumption, and availability.

5.2 GLOBAL OVERVIEW OF AGRICULTURAL WATER SECURITY

5.2.1 GLOBAL AGRICULTURAL WATER USE

Over recent decades, agricultural water use has undergone substantial change (Wu et al., 2022). Numerous earlier and more recent studies (e.g., Postel et al., 1996; Shiklomanov & Rodda, 2003; Foley et al., 2011; Abbott et al., 2019) have consistently shown that agriculture is the largest consumer of global freshwater (blue water), typically accounting for about 70% of total withdrawals. However, some studies, such as Döll and Siebert (2002) and Kummu et al. (2016), have reported even higher estimates, up to 90%. These variations reflect both the complexity of global water use and the limitations of some estimation methods. To address these issues, Wu et al. (2022) developed a new approach using spatially detailed Earth observation data rather than conventional hydrological models. Their findings confirmed that agriculture continues to dominate global water use, accounting for 87% of total water consumption and approximately 60% of freshwater withdrawals in 2015, a noticeable decline from previous common estimates of 70%.

Most agricultural water is used in irrigated croplands and livestock farming, both of which have expanded significantly in recent decades to meet the demands of a growing global population:

- Mehta et al. (2024) reported that the total irrigated area worldwide increased from 297 Mha in 2000 to 330 Mha in 2017, representing an expansion of approximately 11%. Notably, around 50% of this increase occurred in regions already experiencing water stress, such as the Middle East, North Africa, and South Asia, thereby intensifying pressure on scarce freshwater resources. This expansion has increased global reliance on blue water resources, leading to severe over-extraction from major rivers, including the Colorado, Rio Grande-Bravo, and Yellow, as well as from large aquifer systems (Jasechko et al., 2024; Richter, 2025). For instance, Jasechko et al. (2024) found that nearly one-third of the world's regional aquifers have undergone accelerating groundwater level declines over the past 40 years.

- Livestock farming is also an important user of blue water resources, despite its strong dependence on green water (Heinke et al., 2020). Earlier studies, such as Steinfeld et al. (2006), reported that this sector uses about 15% of the blue water consumed by irrigated crops, since these crops are grown using surface or groundwater to produce animal feed. In contrast, it accounts for 33% of the green water used by rainfed crops and 68% of the green water used by permanent pastures and rangelands, where animals typically graze. However, a more recent study by Heinke et al. (2020) quantified that global livestock production annually uses around 4387 km³ of water, of which 94% (4123 km³) is green water and only 6% is blue water. This updated assessment reinforces earlier findings by confirming that livestock systems are overwhelmingly dependent on green water, particularly through rainfed feed production and grazing.

5.2.2 GLOBAL AGRICULTURAL WATER CONSUMPTION

Agricultural water consumption is a valuable indicator of water management and use in agriculture, as it reflects the amount of water that is actually consumed by crops (Boser et al., 2024), primarily through evapotranspiration, the combined processes of soil evaporation and plant transpiration, which is not returned to the environment. This concept is distinct from total water *use*, and the distinction is particularly important for assessing the sustainability of agricultural water management, especially in water-scarce regions. In irrigated agriculture, water consumption is strongly influenced by the efficiency of irrigation systems and the type of crop cultivated (Boser et al., 2024).

Numerous researchers have estimated agricultural water consumption at the global level over various time periods. For instance, over the period 1971–2000, Rost et al. (2008) estimated that global cropland consumed approximately 7200 km³ year^{-1} of green water, representing between 85% and 92% of total crop water consumption, depending on the estimation method. On irrigated cropland, green water accounted for about 20–35% of water consumption. In contrast, blue water consumption was significant only in intensively irrigated regions and was estimated globally at between 636 and 1364 km³ year^{-1}, depending on the method used, highlighting that global green water consumption by crops greatly surpasses blue water consumption.

5.2.3 GLOBAL AGRICULTURAL WATER SCARCITY

Agricultural water scarcity, particularly in its physical form, is widely recognized as one of the most critical global challenges of the 21st century (Jury & Vaux, 2005). This form of scarcity frequently stems from the complex interplay between limited precipitation and a range of socio-economic and environmental factors, including population growth, inadequate economic capacity to access water, and the impacts of climate change (Zisopoulou & Panagoulia, 2021). In this context, Rosa et al. (2020) highlighted that total agricultural water insecurity arises when green water scarcity (GWS), blue water scarcity (BWS), and economic constraints coexist.

TABLE 5.1

Examples of Studies Analyzing Global Agricultural Water Scarcity

Study	Proposed Index	Main Finding
Rosa et al. (2020)	Economic Agricultural Water Scarcity Index	25% of croplands face economic water scarcity
Liu et al. (2022a)	Agricultural Water Scarcity Index	22% of river basins show water scarcity
Liu et al. (2022b)	Agricultural Water Scarcity Index	40% of croplands face water scarcity

In recent years, substantial research efforts have been directed toward incorporating these multiple dimensions into comprehensive assessments of agricultural water scarcity. These efforts have led to notable advancements in how water scarcity is conceptualized and represented, including the introduction of more integrated definitions and indices that account for water use, availability, and access. A selection of these recent contributions is summarized in Table 5.1.

- Rosa et al. (2020) proposed an economic agricultural water scarcity index (WSI_e) for croplands, which focuses on the lack of irrigation caused by limited institutional and economic capacity, rather than hydrological limitations. Using this index, they demonstrated that approximately 25% of global cropland has already been impacted by economic water scarcity.
- In their study, Liu et al. (2022a) introduced the Agricultural Water Scarcity Index (WSI_{ag}), an integrated metric that combines both blue and green water availability to evaluate agricultural water scarcity. The index was validated by comparing its performance with that of a conventional blue-water-based index over the period 1971–2010. The study found that 22% of river basins were classified as water scarce using WSI_{ag}, whereas only 8% were identified as such when using the blue water scarcity index (WSI_{BW}), emphasizing the critical importance of including green water in comprehensive assessments of agricultural water security. Green water should be carefully assessed and given proper attention due to its distinct and essential roles (Figure 5.1).

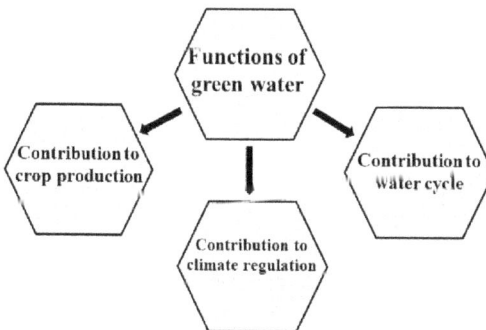

FIGURE 5.1 Schematic representation of the main functions of green water.

These roles include: (1) Supporting crop production by providing soil moisture necessary for plant growth; (2) influencing climate regulation through
evapotranspiration processes, which are crucial for carbon sequestration,
greenhouse gas control, and the maintenance of the planetary energy balance; and (3) regulating the water cycle through evaporation, which impacts
overland water flow and atmospheric interactions (van der Ent et al., 2014).
These roles make green water a critical element in water, soil, and environmental management.

- Liu et al. (2022b) conducted an exemplary study that combines both green
  and blue water components into a single index to assess global agricultural
  water scarcity under both past and future climate scenarios. Specifically,
  they used the validated WSIag to evaluate the extent to which water availability meets agricultural demand. Their analysis, covering both the historical period (1981–2005) and the projected future (2026–2050), revealed that
  approximately 40% of global croplands have already experienced water scarcity. Looking ahead, WSIag projections indicate increased agricultural water
  scarcity in 83%–84% of global croplands compared to the baseline period.

The findings of the aforementioned studies highlight the urgent need for a shift
toward more sustainable agricultural water management practices to ensure long-
term water and food security. This shift is particularly critical given the anticipated
impacts of climate change, which are expected to exacerbate existing challenges
by further intensifying water scarcity. Therefore, immediate action is necessary to
adapt current water management strategies, mitigate these effects, and enhance resilience against future uncertainties.

5.3   REGIONAL OVERVIEW OF AGRICULTURAL WATER SECURITY

5.3.1   REGIONAL AGRICULTURAL WATER USE AND CONSUMPTION

Agricultural water use varies significantly across regions, driven by differences in
hydro-climatic conditions, water availability, crop types, economic development levels, and farming systems, whether irrigated or rainfed.

5.3.1.1   Water Use and Consumption in Asia

In the Asia-Pacific region, the most densely populated region in the world (home to
4.8 billion people), intensive agricultural production, particularly of water-intensive
crops such as rice, sugarcane, and cotton, drives exceptionally high-water use and
consumption. Approximately 90% of freshwater resources (blue water) are allocated
to agriculture, significantly exceeding the global average of around 70% (FAO,
2011). An even higher proportion, about 92%, has been reported by Zhang et al.
(2024) for Southeast Asia, particularly in countries such as Myanmar, Thailand,
Laos, Cambodia, and Vietnam. This heavy dependence on freshwater for farming
exerts considerable pressure on water resources, especially in countries like India,
China, and Pakistan, where agricultural production must meet the demands of large
and growing populations.

Furthermore, the low efficiency of irrigation water use in many Asian countries, for example, 38% in India (Jain et al., 2019), has exacerbated water scarcity challenges. Inefficient irrigation practices not only result in substantial water losses but also contribute to the over-extraction of groundwater, deterioration of water quality, and degradation of agricultural soils through processes such as salinization. Collectively, these issues pose serious threats to long-term agricultural sustainability and regional water security. For instance, approximately 47.5% of irrigated areas in Central Asia are already affected by salinization, resulting in declining crop yields (Hamidov et al., 2016).

5.3.1.2   Water Use and Consumption in Africa

In Africa, agricultural water use is influenced by a combination of arid and semi-arid climates, limited water infrastructure, rapid population growth, uneven water availability, and the prevalence of rainfed farming systems, particularly in sub-Saharan Africa (McClain, 2013; Tuyishimire et al., 2022). On average, agriculture accounts for about 90% of total freshwater withdrawals across the continent (Tuyishimire et al., 2022). Due to the heavy reliance on rainfall (green water), only a small portion of available blue water is used for irrigation in sub-Saharan Africa. In contrast, in North Africa, where rainfall is scarce, there is a higher dependence on blue water for agricultural activities (McClain, 2013).

Tuyishimire et al. (2022) analyzed the water footprint (WF) of food consumption, an important indicator of agricultural water use, in 23 African countries from 2000 to 2018. They found that the WF nearly doubled during this period, rising from approximately 609 km³ to 1212 km³, with an annual increasing rate of 3.7%. Population growth was identified as the major factor of this increase. These findings highlight the growing pressure on water resources in Africa and emphasize the urgent need for sustainable water management strategies to support both rainfed and irrigated farming systems across the continent.

Africa has only 9% of the world's total freshwater resources, compared to 45% in America, 28% in Asia, and 15.5% in Europe (Mugagga & Nabaasa, 2016). As a result, effective water use strategies in agriculture are crucial for improving water and food security. These strategies are also key to achieving several Sustainable Development Goals (SDGs), especially those focused on eliminating poverty (SDG 1), addressing hunger (SDG 2), and ensuring clean water and sanitation (SDG 6) (Mugagga & Nabaasa, 2016).

5.3.1.3   Water Use and Consumption in Europe

On average, agriculture represents approximately 70% of total water use in the European Union (EU) (Gerveni et al., 2020). However, this percentage is significantly higher in certain countries, such as Spain, where agriculture is highly productive, reaching around 82% of total water withdrawals (Serrano et al., 2024). Notably, advancements in irrigation technology and policy-driven water management over recent decades have led to more efficient water use overall, although considerable regional disparities persist. For instance, a study by Gerveni et al. (2020) at the EU level reported that between 1995 and 2010, the largest crop-producing countries experienced an increase in water use, primarily driven by technological changes. In contrast, several Mediterranean countries, affected by water scarcity for years due to higher temperatures and lower precipitation, have managed to reduce their water consumption, largely owing to improvements in water-use efficiency.

5.3.1.4 Water Use and Consumption in North America

Agriculture is the largest consumer of water in many parts of North America, especially in the western USA and northern Mexico, where irrigation is essential due to arid climates. In the USA, the cultivation of many water-intensive crops contributes to approximately 42% of freshwater withdrawals being used for agricultural irrigation (Warziniack et al., 2022), while in Mexico, the figure is approximately 76% (Ochoa-Noriega et al., 2020). In Mexico, inefficient water use, particularly in irrigated areas, is estimated to result in up to 40% water loss (Ochoa-Noriega et al., 2020). In contrast, the situation in Canada is quite different, as 99% of agricultural land relies on precipitation (Biswas, 2003). However, it is worth noting that about 8% of total water withdrawals are still used for agricultural activities, particularly for livestock production, frost control (notably in Ontario and the Maritimes), and crop irrigation in the southern parts of Alberta, British Columbia, Saskatchewan, and Manitoba, to mitigate specific local agricultural challenges such as short drought periods and crop-specific water demands (Biswas, 2003).

5.3.1.5 Water Use and Consumption in Australia

Agriculture remains the dominant sector in terms of water use in Australia, accounting for around 74% of the nation's total water extractions (Zyngier et al., 2024). Of this, approximately 92% is utilized for irrigation purposes (Kelly et al., 2019). A significant portion of irrigated agriculture, over two-thirds, is concentrated in the Murray–Darling Basin (MDB), an extensive system of rivers and lakes covering more than 1,000,000 Km^2 (Shrestha et al., 2023). The primary consumers of irrigation water include crops such as cotton and rice, dairy production, and pastures used for livestock grazing (e.g., Shrestha et al., 2023).

Agricultural water use in Australia is highly variable, primarily driven by fluctuations in seasonal rainfall and the occurrence of droughts (Howden et al., 2014). In response to increasing water scarcity and the growing impacts of climate change, the agricultural sector has increasingly adopted advanced irrigation technologies and enhanced water management practices to reduce water consumption while maintaining or improving crop yields (Howden et al., 2014). For instance, between 2000 and 2010, improvements in water management resulted in a 40% increase in the water-use productivity of irrigated cotton production, contributing to significant yield gains (Roth et al., 2013).

5.3.2 REGIONAL AGRICULTURAL WATER SCARCITY

Agricultural water scarcity, both physical and economic, is a growing challenge worldwide (Mancosu et al., 2015; Rosa et al., 2020). Physical scarcity arises when the demand for green and blue water exceeds available resources (Savenije, 2000), while economic scarcity occurs when renewable blue water exists but cannot be effectively used due to limited financial or institutional capacity (Rosa et al., 2020). This issue is particularly severe in arid and semi-arid regions, where water demand surpasses supply, extreme weather is more common, and population growth intensifies pressure on resources (Stringer & Mirzabaev, 2021).

The following examples highlight regional trends in agricultural water scarcity across Asia, Africa, Europe, North America, and Australia.

5.3.2.1 Water Scarcity in Asia

Agricultural water scarcity in Asia is an escalating concern, driven by the ongoing impacts of climate change, population growth, inefficient irrigation, and increasing competition from urban and industrial sectors. However, the nature of these challenges varies across regions. In South Asia, agricultural areas face acute water shortages and severe groundwater depletion due to overextraction (Lutz et al., 2022). For example, in northwestern India, a major hotspot of water stress, groundwater depletion between 2002 and 2010 was estimated at -24 to -32 km³ (Joshi et al., 2021). Increasing variability in annual rainfall and runoff further intensifies reliance on meltwater and groundwater to compensate for reduced surface water availability during dry periods (Lutz et al., 2022). In Central Asia, farmers primarily contend with economic water scarcity, stemming from limited access to irrigation infrastructure and financial resources (Rosa et al., 2020). Despite relatively abundant water resources, East and Southeast Asia have emerged as global hotspots for water insecurity due to rapid population growth, urbanization, seasonal variability, uneven distribution, and pollution, all contributing to a decline in freshwater availability (Lee et al., 2024). Moreover, Southeast Asia is increasingly affected by the economic consequences of seasonal droughts (Lee et al., 2024). These pressures collectively threaten both water and food security, underscoring the urgent need for improved agricultural water management across the region.

5.3.2.2 Water Scarcity in Africa

In contrast to North African countries such as Tunisia, Algeria, and Libya, where limited rainfall leads to less than 500 m³ of renewable water per capita annually, significantly below the FAO's threshold of 1700 m³ and indicative of acute physical water scarcity and hydrological constraints (Tuyishimire et al., 2022), many sub-Saharan African countries face a different challenge: Economic water scarcity in agriculture. Despite having sufficient water resources to support crop production, these countries often lack the financial investment, institutional support, or infrastructure needed to develop and operate irrigation systems capable of meeting crop water requirements (Rosa et al., 2020). For many of these countries, the installation and expansion of irrigation systems remain financially out of reach, with investment costs reaching up to US$18,000 per hectare (Rosegrant, 1997; Biazin et al , 2012). This highlights the importance of strategic planning to make irrigation more affordable and to improve the efficiency and benefits of rainwater use (Biazin et al., 2012).

5.3.2.3 Water Scarcity in Europe

While most European regions currently have adequate water resources, water scarcity has become a recurring issue in specific areas, driven by rising temperatures, increased evapotranspiration, and shifts in precipitation patterns and river flows

(EEA, 2018). Agricultural water scarcity is a growing concern, particularly in southern and eastern Europe, where climate change has intensified the frequency and severity of droughts during key growing periods, resulting in significant crop yield reductions (EEA, 2018). For example, in 2022, Spain, the world's leading olive oil producer, suffered a drought-induced decline of approximately 50% in olive production compared to the previous season (Wang et al., 2024). Countries such as Spain, Italy, and Greece are under mounting water stress due to their dependence on irrigation and frequent exposure to prolonged dry spells, leading to recurring shortages and reduced water availability for agriculture (e.g., García-Ruiz et al., 2011).

Although economic agricultural water scarcity is generally uncommon across most of Europe, Rosa et al. (2020) found that certain regions, particularly in Eastern Europe, still face this issue. Key contributing factors include inadequate management of blue water resources, weak policy implementation, low economic feasibility of irrigation to offset green water deficits, and limited access to efficient irrigation technologies.

In response to these challenges, the EU has implemented several policy measures aimed at promoting sustainable water use, notably through the Common Agricultural Policy (CAP) and the Water Framework Directive (WFD). These initiatives encourage the adoption of advanced irrigation systems, the application of water pricing instruments, and the reuse of treated wastewater to optimize water efficiency and reduce pressure on freshwater supplies (e.g., Hristov et al., 2021). Collectively, these strategies aim to enhance the resilience of European agriculture and ensure its long-term sustainability under increasing water stress.

5.3.2.4 Water Scarcity in North America

In many parts of North America, especially the American Southwest and northwestern Mexico, water scarcity has become an increasingly pressing issue over the past few decades. This is due to the frequent occurrence of extreme weather events, such as droughts and heatwaves, changes in precipitation patterns, as well as non-climatic factors including population growth and changing consumption patterns that drive rising water demand (Yáñez-Arancibia & Day, 2017).

5.3.2.5 Water Scarcity in Australia

Similar to many arid regions worldwide, water resources allocated for farming in Australia, particularly in the MDB, have significantly declined due to prolonged dry periods, such as the record-breaking Millennium Drought from 1997 to 2009, and reduced cool-season rainfall (Prosser et al., 2021). These conditions have altered key hydrological processes, including weakened surface–groundwater connections and increased evapotranspiration (Prosser et al., 2021). Consequently, the irrigated area within the basin has decreased, resulting in a notable decline in agricultural production. For example, between 2001 and 2006, drought and related water scarcity caused a significant drop in irrigated crop yields; rice production, for example, reduced by 643 kt, decreasing Australia's contribution to both local and global food security, given the region's role as a key food production zone (Qureshi et al., 2013).

5.4 LOCAL AGRICULTURAL WATER SECURITY

Assessing agricultural water security at the local scale is crucial for informed decision-making. Such assessments help identify key influencing factors, such as the spatial distribution of precipitation, hotspots of extreme weather events, land use patterns, farming practices, and pollution issues, which often vary significantly within a given area (Dang et al., 2021). Recognizing these local dynamics enables the development of targeted and efficient water management strategies, which can enhance water availability for farmers and support the long-term sustainability of agricultural systems (Mehla, 2022). The following examples present findings from studies that have examined agricultural water security at both national and local scales.

5.4.1 EXAMPLES FROM ASIA

Numerous studies have addressed local agricultural water scarcity in Asia. The following are selected case studies that illustrate different issues and challenges across the region (Table 5.2).

5.4.1.1 China

A study by Huang et al. (2019b) on agricultural water security in North China, a major grain-producing region (23% of national output), is a good illustrating example of how non-climatic factors can significantly reduce water availability. In other words, the study showed that although precipitation declined only slightly (by 1.9–15.6%) between 1998 and 2015 compared to 1956–2000, rapid economic growth and urbanization, along with associated changes in land use and land cover, led to a sharp decline in renewable water resources: Surface water dropped by about 16% and shallow groundwater by 24%.

5.4.1.2 Bangladesh

Considering groundwater depth, river water availability, rainfall, and the salinity of irrigation water (from both surface and groundwater resources) as key indicators for

---

TABLE 5.2

Examples of Studies on Agricultural Water Security in Selected Asian Countries

Study	Country	Contributors to Water Challenges
Huang et al. (2019b)	China	Economic growth, urbanization, and land use change
Ahammed et al. (2018)	Bangladesh	Salinisation and limited availability
Mehla (2022)	India	High consumption of agricultural water
Janjua et al. (2021)	Pakistan	Droughts, water overuse, and corruption
Ahmad and Al-Ghouti (2020); Aloui et al. (2023)	Qatar	Groundwater overextraction and salinity

assessing agricultural water scarcity in Bangladesh, Ahammed et al. (2018) found that approximately 19% of the country's land area faces a high risk of water scarcity. The study identified that, spatially, in the northwest, the primary driver is limited water availability, whereas in the southwest, the issue is more closely linked to deteriorating water quality, particularly salinity problems caused by seawater intrusion into coastal aquifers. These findings highlight the urgent need to implement effective measures to prevent further degradation; otherwise, water scarcity is likely to intensify in the future.

5.4.1.3   India

Mehla (2022) assessed the WF of the semi-arid Banas River Basin (BRB) in India over the period 2008–2020 and found that approximately 95.5% of the total WF (19.3 out of 20.2 billion m³) was attributed to agriculture. This underscores the urgent need to improve water productivity and adopt more sustainable agricultural practices to safeguard water security in the basin's agricultural areas and meet the growing demand for food for local people. Notably, the study revealed that four crops, wheat, bajra, maize, and rapeseed and mustard, accounted for about 67% of the estimated agricultural WF.

5.4.1.4   Pakistan

The Indus Basin in Pakistan exemplifies how climatic and non-climatic challenges can critically threaten the agricultural water security of the whole country, as it provides essential irrigation, supporting nearly 90% of Pakistan's food production and sustaining about 80% of its population (Janjua et al., 2021). Janjua et al. (2021) offer an excellent analysis of these challenges, highlighting several key issues: (1) Droughts that are causing significant water shortages; (2) non-climatic factors such as urbanization, rapid population growth, and irrigation water losses (e.g., canal seepage, field application losses), which are worsening water deficits; (3) overextraction that is leading to declining groundwater levels and the salinization of approximately 2.25 million hectares of irrigated land; and (4) corruption, which may be hindering any plans for effective water management.

5.4.1.5   Qatar

In Qatar, farming and other human activities are heavily dependent on groundwater due to the country's hyper-arid climate. However, over the past few decades, significant changes in the groundwater regime, both in terms of quantity and quality, have raised serious concerns about water security for farmers and other users. In terms of quantity, Aloui et al. (2023) reported that the current rate of groundwater extraction is approximately 250 million m³ year^{-1}, which is five times greater than the sustainable limit. Regarding quality, Ahmad and Al-Ghouti (2020) noted that Qatar's groundwater is predominantly brackish to saline. In 1992, only 8% of the country's wells contained non-saline water, with salinity levels below 0.7 dS m^{-1}; however, by 2012, this category had completely disappeared (Ahmad & Al-Ghouti, 2020; Aloui et al., 2023). These issues highlight the urgent need to consider alternative water resources (Ajjur & Al-Ghamdi, 2022).

5.4.2 Examples from Africa

Several studies have investigated agricultural water scarcity at the local level across Africa. The following selected case studies highlight the main water challenges posing risks to this vulnerable continent (Table 5.3).

5.4.2.1 Nigeria

Although Nigeria possesses abundant water resources, both surface and groundwater, pollution and various non-climatic factors severely limit the effective use of these resources, resulting in a critical water crisis and agricultural water insecurity. In other words, despite the abundance of freshwater, the country currently faces a serious water crisis due to the degradation of groundwater and surface water quality caused by industrial waste, agricultural runoff, domestic sewage, and flooding (Isukuru et al., 2024). Furthermore, desertification, particularly in the northern regions, along with an increasing number of illegal water wells, rapid population growth, accelerated urbanization, agricultural intensification, inadequate water governance, insufficient water infrastructure, and corruption, further exacerbate the situation (Isukuru et al., 2024). Collectively, these challenges make resolving the water crisis highly complex, highlighting the urgent need for long-term strategies and efforts (Isukuru et al., 2024).

5.4.2.2 Cameroon

Cameroon is another illustrative example in Africa where, despite the abundance of both surface and groundwater resources, a growing number of farmers, particularly women, face physical and economic water scarcity due to the effects of climate change, such as erratic rainfall and temperature fluctuations, and weak water governance (Fonjong & Zama, 2023). In many parts of the country, agricultural water insecurity goes beyond these forms of scarcity, further restricting the ability of farmers, especially women, to carry out their varied responsibilities within predominantly agricultural societies (Fonjong & Zama, 2023).

5.4.2.3 South Africa

Droughts in Africa can critically affect agriculture, particularly because the majority of farming systems are rainfed and rely heavily on seasonal rainfall. Any reduction

---

TABLE 5.3
Examples of Studies on Agricultural Water Security in Selected African Countries

Study	Country	Contributors to Water Challenges
Isukuru et al. (2024)	Nigeria	Pollution, population growth, poor governance, corruption
Fonjong and Zama (2023)	Cameroon	Erratic rainfall, weak governance, water scarcity
Meza et al. (2021)	South Africa	Droughts, land degradation, poor water storage
Musonda et al. (2020)	Zambia	Frequent droughts
Hamed et al. (2018)	Tunisia	Frequent droughts, surface water decline, groundwater salinity

in rainfall during dry periods significantly lowers green water availability, leading to significant decreases in crop yields. South Africa is a clear example of this challenge. A local-level study by Meza et al. (2021) found that all municipalities in the country have experienced drought over the past 30 years. The study highlighted that rainfed farming systems are especially vulnerable in the northern, central, and western regions. These findings underscore the urgent need for local strategies to reduce drought risk, including efforts to prevent land degradation and enhance water storage infrastructure.

5.4.2.4   Zambia

Zambia is one of the African countries most affected by prolonged drought, resulting in severe water scarcity and serious consequences for agriculture and rural livelihoods. A study by Musonda et al. (2020) reported moderate to severe droughts in 1991–1992, 1994–1995, 2004–2005, and 2015–2016. Using the Standardized Precipitation Index (SPI), the study also revealed a clear upward trend in drought frequency and intensity over the past 30 years. Regionally, the study found that droughts were most intense, prolonged, and severe in the southwest, while only moderate conditions were recorded in limited areas of the northeast. These findings underscore the urgent need for strategic adaptation and mitigation measures to manage the increasing impacts of drought and to enhance agricultural water security.

5.4.2.5   Tunisia

Tunisia, a North African country, faces severe agricultural water insecurity. Over the past decade, climate change has increasingly affected the country's water resources, both in quantity and quality. Recurring droughts have led to a decline in surface water availability and a noticeable deterioration in groundwater systems, particularly due to rising salinity levels (Hamed et al., 2018). These challenges have left many farmers without adequate water, especially during the critical growth stages of crops, putting agricultural production at serious risk.

5.4.3   EXAMPLES FROM EUROPE

Even though most European countries are considered water-rich, climatic conditions such as droughts and floods over the past few decades have placed these resources under increasing pressure, leading to serious concerns about agricultural water security. The following selected case studies highlight some of these concerns (Table 5.4).

5.4.3.1   Germany

Germany is among the European countries increasingly affected by the frequent occurrence of droughts and drier summers in recent years (EEA, 2018). These climatic changes have had significant impacts on green and blue water availability, water extraction, crop water requirements, irrigated areas, the water retention capacity of cultivated soils, and crop yields. For example, a study by McNamara et al. (2024), which examined all crop areas in Germany, found that irrigation requirements during droughts increased by an average of 72% compared to non-drought periods. The study further showed that this increase was more pronounced for crops

TABLE 5.4

Examples of Studies on Agricultural Water Security in Selected European Countries

Study	Country	Contributors to Water Challenges
McNamara et al. (2024)	Germany	Droughts, increased irrigation demand
Vasilakou et al. (2025)	Greece	Droughts, groundwater overexploitation, poor governance
Yimer et al. (2023)	Belgium	Agricultural drainage
Luo et al. (2024b)	Italy	Saltwater intrusion, increasing salinity
Lovrinović et al. (2023)	Croatia	Seawater intrusion, groundwater salinity

cultivated on silty soils than on sandy soils. Overall, these findings highlight the increasing pressures on agricultural water security, emphasizing the importance of adopting optimal irrigation strategies and technologies at the farm level that promote the conservation of water resources.

5.4.3.2 Greece

Greece is a notable example of a European country where agricultural water security is increasingly at risk. This vulnerability is driven by the frequent occurrence of droughts, recorded four times over the period 1990–2010, in addition to the ongoing drought, combined with the overexploitation of groundwater resources, inefficient surface runoff management, and overall poor water governance (e.g., Vasilakou et al., 2025). These challenges have led to recurring water shortages, underscoring the urgent need for more effective water resource management to ensure long-term water security.

5.4.3.3 Belgium

In many parts of Europe, drainage within agricultural plots is essential to prevent issues such as waterlogging and salinization, and to maintain optimal soil moisture and aeration around crop roots. These conditions are important for optimal crop growth and high yields. Due to these benefits, it is estimated that approximately 33% of northwestern Europe is drained (Gramlich et al., 2018). However, it is worth noting that extensive drainage in some areas has affected catchment hydrology by reducing groundwater volumes, lowering water table levels, and accelerating the transport of pollutants to streams (Hasselquist et al., 2018). The Kleine Nete watershed in Belgium is one such example. Using a coupled hydrological modelling tool, SWAT+gwflow, Yimer et al. (2023) found that intensive drainage led to a 50% reduction in groundwater saturation excess flow, an 18.4 mm increase in drainage discharge to streams, and a 0.12 m drop in the basin's average water table depth. These results underscore the need for strategies for the long-term preservation of the watershed.

5.4.3.4 Italy

In Italy, saltwater intrusion is becoming an increasingly serious challenge, affecting vast areas of agricultural land. One of the most impacted regions is the Po River Delta, an important river system and agricultural zone. Over the past few decades, this issue, exacerbated by more frequent and intense droughts, has led to critical problems such as rising salinity levels in both water and soil, as well as a significant decline in freshwater availability for farmers (Luo et al., 2024b). These developments underscore the urgent need for effective adaptation measures to enhance water security in the region for farmers and other users.

5.4.3.5 Croatia

Seawater intrusion is also a significant issue in many agricultural coastal areas of Croatia, leading to numerous problems, particularly increased groundwater salinity. This rise in salinity can reduce the availability of freshwater resources of adequate quality for farming practices. A study conducted by Lovrinović et al. (2023) on the Neretva Valley coastal aquifer system showed that during dry periods, the groundwater regime is primarily influenced by seawater intrusion from both the sea and the Neretva River bed. Conversely, during rainy periods, increased river discharge helps flush seawater out of the riverbed, thereby improving groundwater quality. Understanding this local dynamic of seawater intrusion is essential for developing adaptation measures that can ensure the long-term protection of water resources from salinity-related issues.

5.4.4 EXAMPLES FROM NORTH AMERICA

In North America, spatial and temporal variability in water availability, alongside rising demand and the impacts of climate change, have prompted numerous researchers to assess the current state of agricultural water security. The following case studies illustrate the key challenges facing agricultural water security in this region (Table 5.5).

5.4.4.1 Canada

One of the most direct impacts of global warming in Canada is glacier melting, which significantly affects both surface and groundwater resources. In the short term, this process may enhance agricultural water security. For example, Castellazzi et al. (2019) found that glacier mass loss in the Canadian Rocky Mountains (CRM), averaging approximately −3056 million m^3 per year between 2002 and 2015, contributed between 288 and 1717 million m^3 annually to surrounding watersheds and increased groundwater storage by 3976 million m^3 per year. These findings suggest

---

TABLE 5.5

Examples of Studies on Agricultural Water Security in Selected North American Countries

Study	Country	Contributors to Water Challenges
Castellazzi et al. (2019)	Canada	Glacier mass loss
Jones and van Vliet (2018)	USA	Droughts, water scarcity, increased salinity
Arreguín-Cortéz et al. (2011)	Mexico	Water scarcity, pollution, aquifer overexploitation

that an important portion of the meltwater is retained in aquifers rather than flowing directly into rivers. However, further research is needed to assess the long-term implications of this melting on water resources.

5.4.4.2 USA

Several parts of the USA, particularly the southern regions, have experienced frequent droughts over the past few decades. The consequences of these droughts on water resources extend not only to reduced availability and increased scarcity but also to the degradation of water quality, such as elevated salinity levels. For instance, a study by Jones and van Vliet (2018), which focused on the southern USA, found that drought conditions led to a 21% increase in river water salinity compared to non-drought periods. The study further revealed that water scarcity was intensified by drought-related declines in supply, rising sectoral demands, and deteriorating water quality. Together, these findings underscore the need to consider both quantity and quality dimensions when assessing the impacts of drought on water resources.

5.4.4.3 Mexico

Agricultural water security in Mexico is in a critical state, as the country faces a range of challenges including water scarcity, declining water quality, overexploitation of aquifers, poor governance of available freshwater resources, the increasing impacts of climate change, and growing competition between agricultural, industrial, and domestic water demands (Arreguín-Cortéz et al., 2011). These issues collectively highlight the urgent need to develop long-term strategies that can help to manage water resources sustainably and efficiently.

5.4.5 EXAMPLES FROM AUSTRALIA

Most parts of Australia have experienced an increased frequency and intensity of droughts over the last few decades, raising serious concerns about the availability of water for farming activities. Agricultural drought, characterized by prolonged periods of insufficient soil moisture, is one such type of drought that poses significant risks to crop production and farm sustainability. Specifically, a study by Hoque et al. (2021), conducted in Northern New South Wales using geospatial methods, found that approximately 19% of the studied region is at very high risk, while about 41% is at moderate to high risk of agricultural drought. These findings emphasize the urgent need for effective drought mitigation and water management strategies to safeguard agricultural productivity and water security in the region. Intensified agricultural droughts may lead to a critical decline in food production.

5.5 CONCLUSIONS

The findings of this chapter clearly show that agricultural water security is critically threatened at local, regional, and global levels due to a combination of climatic factors, such as climate change, and non-climatic drivers including population growth, urbanization, and land use changes. This underscores the urgent need for enhanced conservation of limited freshwater resources. The case studies presented from various regions highlight significant differences in water availability, climate impacts, and water management options on farms, emphasizing the importance of region-specific strategies. Moreover, agricultural water security must be assessed by considering both water quantity and quality. In conclusion, understanding current trends in agricultural water security is an essential first step toward developing effective adaptation strategies and should be complemented by projections of future trends to adequately plan the necessary actions to ensure the sustainability of agricultural production in an increasingly water-scarce world.

6 Future Trends in Agricultural Water Security under Climate Change
Global, Regional, and Local Perspectives

6.1 INTRODUCTION

Globally, agriculture is the dominant water consumer, accounting for approximately 70% of total water withdrawals (e.g., Foley et al., 2011; Abbott et al., 2019). This high percentage is primarily driven by a combination of biophysical, climatic, human, and water management-related factors. First, physiological processes such as transpiration, photosynthesis, respiration, and nutrient transport require substantial amounts of water to support stable and optimal crop growth and yields (McElrone et al., 2013). These amounts are particularly pronounced for water-intensive crops like rice and wheat, which are among the most widely cultivated crops worldwide (Bhatt et al., 2016). Second, in arid and semi-arid regions, high evapotranspiration rates combined with limited rainfall necessitate the application of large volumes of irrigation water to meet crop water requirements (Gong et al., 2025). Third, rapid global population growth, from 5 billion in 1986 to around 7.8 billion by 2020 (Gu et al., 2021), along with rising incomes and dietary shifts towards more water-intensive foods, has further intensified agricultural water use to meet growing food demands (Davis et al., 2017). Finally, it is worth noting that poor soil and water management practices have resulted in significant water losses in agricultural areas, thereby increasing the need for additional water applications (e.g., Mohammadi et al., 2019). According to FAO estimates, about 50% of water is lost through evaporation, seepage, and poor water management (Renault et al., 2013).

As water is the most fundamental driver shaping agricultural production, ensuring its sustainable availability with adequate quality for farmers is vital to achieving food security for a growing population, supporting rural livelihoods, enhancing the agricultural economy, and maintaining ecosystem stability (Amparo-Salcedo et al., 2025). However, a growing number of climatic and non-climatic factors are intensifying pressure on agricultural water resources worldwide, putting their sustainable availability with adequate quality at high risk (Aligholi & Hayati, 2022). This, in

DOI: 10.1201/9781003660521-6

turn, may lead to decreased agricultural production, loss of economic and social stability, and increased food insecurity (Ouda & Zohry, 2020). Also, it may delay progress toward achieving the United Nations Sustainable Development Goals (SDGs), such as Zero Hunger (SDG 2), Clean Water and Sanitation (SDG 6), Good Health (SDG 3), and Life on Land (SDG 15) by 2030 (Taka et al., 2021).

Climate change is one of the most critical factors threatening agricultural water security. Shifts in precipitation patterns, rising temperatures, and the increasing frequency of extreme climate events, such as droughts, heatwaves, and intense rainfall, are significantly altering both the quality and quantity of freshwater resources available for farming, as well as increasing agricultural water requirements. These changes critically affect crop growth and yield potential (Cai et al., 2015). Projections indicate that these impacts will continue in the coming decades, underscoring the urgent need to deepen our understanding of climate change effects in order to address water-related challenges in agriculture and to develop suitable adaptation strategies (Amparo-Salcedo et al., 2025). This need is particularly pressing in regions where agricultural production and socioeconomic conditions are highly vulnerable to climate change, such as in Africa and Asia (Haj-Amor et al., 2023).

In this context, and drawing on findings from recent studies on future projections of agricultural water security, this chapter examines anticipated trends in water quantity, water quality, and agricultural water requirements at global, regional, and local scales. By integrating insights from the latest scientific research and case studies from diverse regions, the chapter aims to identify emerging risks and challenges in the management of water within agricultural systems.

6.2   POSSIBLE EFFECTS OF CLIMATE CHANGE ON AGRICULTURAL WATER RESOURCES

The possible effects of climate change on agricultural water resources have been investigated by several researchers at various spatial scales, ranging from local and regional to global levels (e.g., Fischer et al., 2007; Yan et al., 2018; Rameshwaran et al., 2021; Liu et al., 2022b). Overall, these studies suggest that climate change will likely lead to a decline in agricultural water availability in terms of both quantity and quality, as well as an increase in agricultural water requirements (Figure 6.1), with notable variation among regions.

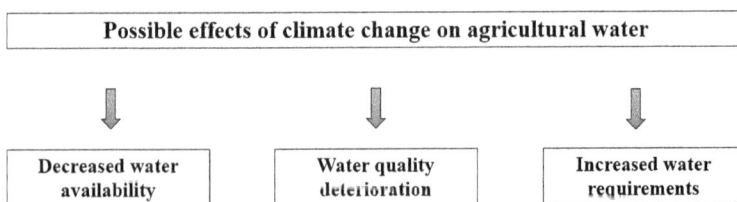

FIGURE 6.1   Possible effects of climate change on agricultural water resources.

6.2.1 EFFECTS ON WATER AVAILABILITY

Climate change is expected to significantly disrupt the hydrological cycle, particularly through shifts in temperature, which intensifies evaporation, and altered rainfall patterns. These changes can strongly influence key hydrological processes such as infiltration, evapotranspiration, runoff, and groundwater recharge, ultimately affecting the distribution and availability of both green and blue water resources (Ehtasham et al., 2024). According to Konapala et al. (2020), increased seasonal variability in rainfall and evaporation may lead to pronounced disruptions in water availability. In regions with a single rainy season, longer and more frequent dry spells are likely to emerge, while in wetter areas, precipitation may become concentrated in shorter periods, raising the risk of flooding and complicating reservoir operations. These changing patterns, combined with more frequent climate extremes such as droughts and heatwaves, are expected to hinder the recharge of water storage systems and intensify agricultural water scarcity in many parts of the world (Liu et al., 2022b).

6.2.2 EFFECTS ON WATER QUALITY

It is not only the quantity of water that will be affected; the quality of available water is also likely to deteriorate due to increased pollution loads, elevated evaporation rates, and reduced dilution capacity, all consequences of climate change impacts such as flooding and droughts (e.g., Mosley, 2015; Asresu et al., 2025). Poor water quality can further limit its use in agriculture, particularly irrigation, especially where salinity, nutrient loading, or contamination becomes a concern under changing climatic conditions.

6.2.3 EFFECTS ON WATER REQUIREMENTS

On the other hand, while a decline in both water quantity and quality is anticipated, most countries are also projected to face increased agricultural water requirements (Fischer et al., 2007; Mancosu et al., 2015). This trend is largely driven by rising temperatures and higher evapotranspiration demands. These challenges underscore the necessity of planning optimal agricultural water management strategies to avoid an imbalance between water demand and availability.

6.3 GLOBAL EFFECTS OF CLIMATE CHANGE

Climate change can alter the global hydrological cycle primarily due to rising temperatures, shifts in precipitation patterns, and an increased frequency of extreme weather events such as droughts and floods (Du et al., 2021). These changes may intensify global water stress issues, including reduced water availability, increased water requirements, and the deterioration of water quality.

6.3.1 EFFECTS ON AGRICULTURAL WATER AVAILABILITY

Since temperature, precipitation, and evaporative demand are the most influential factors affecting water availability (Konapala et al., 2020), and as these factors are

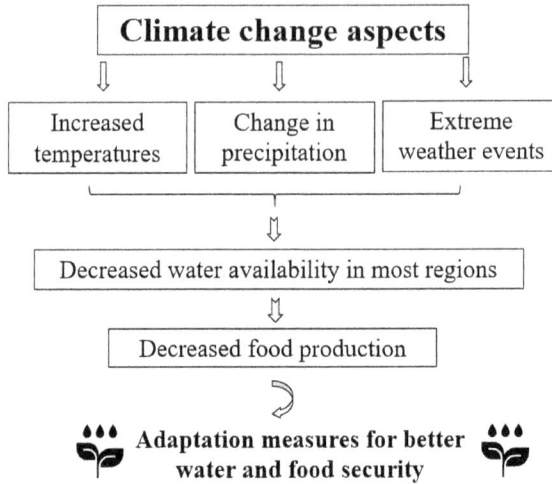

FIGURE 6.2   Effects of climate change on water availability and food production.

strongly affected, and will continue to be affected, by climate change, it is evident that climate change will significantly impact agricultural water availability from soil (i.e., moisture), surface resources (e.g., rivers, lakes), and aquifers. More specifically, increased temperatures, and consequently increased evapotranspiration, along with altered precipitation patterns and extreme weather events such as floods, heatwaves, and droughts, are expected to lead to severe water scarcity, affecting both blue and green water resources in most regions of the world (Figure 6.2). Liu et al. (2022b) analyzed agricultural water scarcity, considering both green and blue water components, at the global level for a historical period (1981–2005) and a future period (2026–2050), under RCP2.6 and RCP6.0 climate change scenarios. They revealed that approximately 40% of global croplands have already experienced water scarcity, and this percentage is projected to increase by more than 3% in the future. They also showed that reduced water availability and increased crop water requirements in most cropland areas are the main factors driving future water scarcity.

The Intergovernmental Panel on Climate Change (IPCC)'s Fifth Assessment Report anticipates that global atmospheric temperatures could rise by as much as 4°C by 2100, a shift likely to significantly affect water availability across the globe. At the same time, more intense rainfall is expected to increase surface runoff, elevate flood risks, and reduce the replenishment of groundwater reserves (Şen, 2021). These projected shifts in temperature and precipitation patterns are likely to intensify the ongoing decline in agricultural water availability in most regions of the world. This potential limitation may pose a serious threat to global food security, especially considering that, by 2050, around 40% of the world's population is expected to live in regions lacking adequate land and water resources to sustain their own food production (Ibarrola-Rivas et al., 2017).

Given that agricultural water availability is a critical requirement for sustaining long-term agricultural production, the anticipated impacts of climate change over the coming decades highlight the urgent need for enhanced efforts towards the sustainable management of existing water resources. Such measures are essential to mitigate the potential adverse effects of future water-related challenges on agricultural productivity and global food security, particularly in the face of a growing population (Figure 6.2).

6.3.2 EFFECTS ON AGRICULTURAL WATER REQUIREMENTS

Investigation of future agricultural water demand under climate change has advanced significantly in recent years. Early studies relied on simplified models and a limited number of climate change scenarios, providing broad estimates of future water needs, particularly irrigation requirements. In contrast, more recent research employs advanced modeling frameworks and a wider range of climate projections, allowing for more precise and region-specific assessments. These contemporary studies integrate numerous variables, including evolving precipitation patterns, rising temperatures, changes in crop types and growing seasons, and the physiological impacts of increased atmospheric CO_2 concentrations. Particular attention is given to future irrigation requirements, which are crucial for water resources planning and management. Examples of these early and more recent efforts are summarized in Table 6.1:

- Döll and Siebert (2002) used the WaterGAP model, together with climate projections from the HadCM2 and ECHAM4 models under the IPCC IS92a scenario, to assess the effects of climate change on net irrigation requirements (NIR) at the global scale. Relative to the reference period 1961–1990, the study projected a global increase in NIR of approximately 5.5% by the 2070s, driven by rising temperatures and altered precipitation patterns. The findings also indicated that South Asia would experience the highest increase in NIR among all regions, underscoring the urgent need to develop adaptation strategies in this region to prevent severe water stress and safeguard food production under changing climatic conditions.

TABLE 6.1
Overview of Key Studies Assessing Climate Change Effects on Global Irrigation Requirements

Study	Model Used	Key Finding
Döll and Siebert (2002)	WaterGAP	Global net irrigation requirements will increase by 5% by 2070
Fischer et al. (2007)	AEZ	Global net irrigation requirements will increase by 30% in developing regions by 2080
Konzmann et al. (2013)	LPJmL	Considerable regional variability in climate change effects on irrigation demand
Wada et al. (2013)	Many models	Irrigation demand will increase, especially in the Northern Hemisphere by 2100
Haile et al. (2024)	H08	Actual irrigation water withdrawals will nearly double by 2100 in several regions

- Fischer et al. (2007), employing the AEZ modeling approach, assessed the impact of the A2r climate change scenario on global NIR. Their analysis placed particular focus on the potential for mitigation strategies to offset rising irrigation demand. Between 2000 and 2080, they projected that NIR would increase by approximately 56% in developing regions and by about 16% in developed regions. The largest relative increases were expected in South Asia, Africa, and Latin America due to increased evapotranspiration and changing rainfall patterns. Notably, the study also found that implementing mitigation measures could reduce the effect of climate change on NIR by roughly 40% compared to a scenario without mitigation measures, highlighting their significance for sustainable water management in agriculture.
- Konzmann et al. (2013) employed the LPJmL model to evaluate the effects of climate change on global irrigation requirements. Relative to the reference period 1971–2000, the study estimated that by the 2080s, global NIR could decline by approximately 17%. This decline was attributed primarily to the positive impacts of elevated atmospheric CO_2 concentrations on crops, shorter growing periods, and increased regional precipitation. However, the study also mentioned that certain regions, such as Southern Europe, parts of Asia, and North America, could face increases in NIR exceeding 20%, emphasizing the considerable regional variability in climate change effects on irrigation demand.
- Wada et al. (2013) used a suite of global hydrological models (GHMs), including H08, LPJmL, MPI-HM, PCR-GLOBWB, VIC, WaterGAP, and WBMplus, to estimate the potential effects of climate change on irrigation water demand (IWD) by 2100. The study found that IWD is likely to increase, with the extent of this rise strongly dependent on the degree of global warming and changes in regional precipitation. Specifically, it showed that under the high-emission scenario (RCP8.5), summer IWD in the Northern Hemisphere could rise by more than 20% by 2100. Furthermore, the study highlighted considerable uncertainty in these estimations, mainly driven by the choice of GHMs, which contributed more to overall variability than the selected global climate models (GCMs) or emission scenarios.
- Haile et al. (2024) used the H08 global hydrological model to assess the effects of SSP370 and SSP585 scenarios on actual irrigation water withdrawals (AIWW). They projected increases of 96 to 107% for the period 2041–2100 relative to 1981–2014, with the largest rises in India, South China, parts of the USA, Europe, South Africa, and Latin America. These findings highlight the urgent need for regional adaptation strategies to mitigate water stress and safeguard food and water security.

In addition to irrigation water, i.e., blue water, aspects of climate change, particularly increased temperatures and altered rainfall patterns, may pose a serious threat to the availability of green water. Mbewu et al. (2024) revealed that climate change can significantly affect the green water footprint. Specifically, they noted that increased rainfall and temperatures can lead to a higher green water footprint, while limited rainfall results in a lower green water footprint.

He and Rosa (2023) found that green water scarcity already affects food production for about 890 million people, with projections indicating this number could rise to 1.23 and 1.45 billion under global warming scenarios of 1.5°C and 3°C, respectively. However, they also indicated that implementing adequate soil moisture conservation measures could reduce this increase, thereby contributing to improved productivity of rainfed agriculture.

Based on the above-mentioned findings, it is clear that climate change poses a threat to both blue and green water resources, highlighting the urgent need to implement adequate adaptive water management strategies to ensure sustainable agricultural water use and global food security.

6.3.3 EFFECTS ON WATER QUALITY

Since water quality, particularly surface water (SW) quality, is strongly influenced by the surrounding climate, it is increasingly recognized that both current and future climate change-related conditions, such as rising temperatures, altered rainfall patterns, and more frequent extreme weather events like droughts and floods, will affect key water quality parameters, including salinity, micronutrient levels, pathogen presence, pH, and dissolved oxygen (Amanullah et al., 2020). For instance, a global review by van Vliet et al. (2023) found that droughts and heatwaves can reduce dissolved oxygen and increase river temperature, salinity, algae growth, and the concentration of harmful substances due to reduced dilution, whereas floods can enhance the transport of plastics, sediments, nutrients, and trace contaminants from runoff, though high flows may dilute salinity and other dissolved materials.

Research has shown that changes in rainfall patterns, temperature variations, and changes in hydrological processes can significantly affect the transport and retention of nutrients in stream systems (e.g., Whitehead et al., 2019). Rising temperatures, when combined with fluctuating nutrient concentrations, may severely impact the availability and quality of water for agricultural purposes (Dorado-Guerra et al., 2023). These climatic stressors, along with increasing anthropogenic nutrient loads, are expected to further deteriorate water quality in the near future (e.g., Dorado-Guerra et al., 2023; Yuan et al., 2023). Consequently, the suitability of agricultural water for essential purposes such as irrigation, livestock watering, and food processing may be increasingly limited (Figure 6.3).

In addition to its critical effect on SW, climate change can also deteriorate the quality of aquifers. A review by Dao et al. (2024) highlighted that climate change-related conditions such as warming trend, intense rainfall, and sea level rise can significantly affect aquifers. First, the warming trend can alter chemical and physical processes within aquifers, impacting those up to 100 m deep. Second, intense rainfall accelerates chemical leaching, increasing concentrations of suspended and dissolved solids. Finally, sea level rise promotes saltwater intrusion, a process by which saltwater penetrates into coastal aquifers, thereby increasing their salinity. As aquifers are essential freshwater resources, especially in arid and semi-arid regions, such deterioration may reduce their suitability for agriculture.

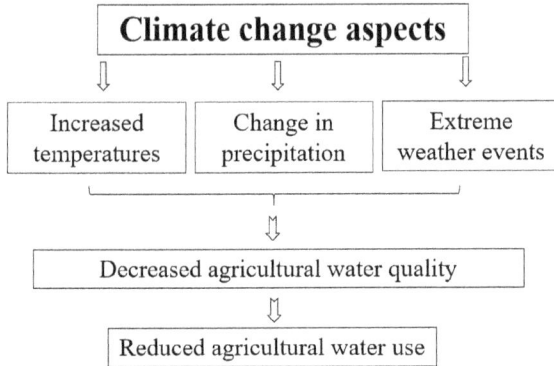

FIGURE 6.3 Effect of climate change on water quality and use.

It is clear that climate change will lead to critical deterioration of water quality in both surface and deep-water resources. This, combined with the expected decline in water quantity, may put agricultural water security at high risk, especially in arid and semi-arid areas. This anticipated situation emphasizes the urgent need to intervene now to prevent further decreases in water quality and to avoid any restrictions on the use of agricultural water for activities such as irrigation, livestock watering, and food processing. Adaptation measures must focus on both water quality and quantity, as both are essential to conserving water.

6.4 REGIONAL EFFECTS OF CLIMATE CHANGE

Climate change is anticipated to lead to major changes in water resource systems across most regions of the world. These changes may manifest in various forms, such as increased water demand, intensified water scarcity and shortage, and reduced water quality, depending on the region. This highlights that each region will face its own specific future challenges concerning agricultural water security. The following sub-sections present these manifestations.

6.4.1 PROJECTIONS FOR ASIA

Asia, home to billions of people, is a region where freshwater resources are intensively used for agriculture and where a remarkable diversity of climates is observed, from the arid regions of West Asia to the monsoon-driven plains of South and Southeast Asia. However, water resources in this region are currently facing, and are expected to face, considerable changes in both quantity and quality due to climate change-related conditions such as increased temperatures, altered precipitation patterns, accelerated glacier melting, and more frequent extreme weather events. These changes pose serious challenges to agricultural water security, particularly in areas that are already water-stressed or highly dependent on seasonal water flows. The following points outline the potential impacts of climate change on agricultural water

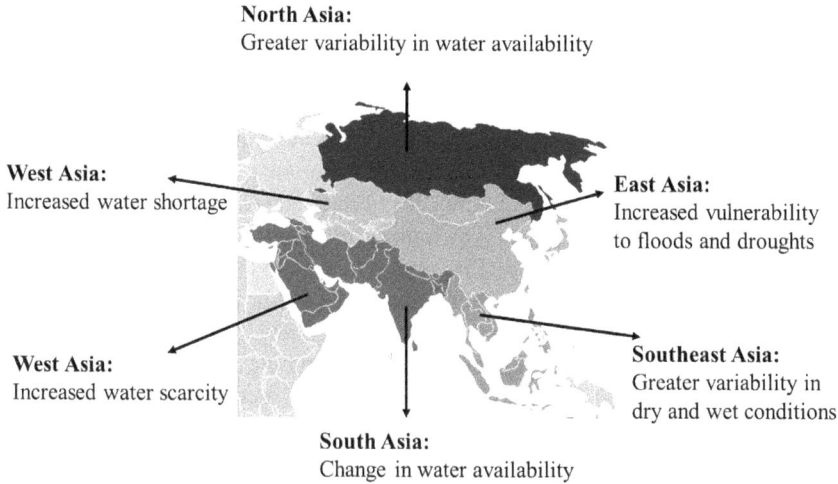

North Asia:
Greater variability in water availability

West Asia:
Increased water shortage

East Asia:
Increased vulnerability
to floods and droughts

West Asia:
Increased water scarcity

Southeast Asia:
Greater variability in
dry and wet conditions

South Asia:
Change in water availability

FIGURE 6.4   Examples of potential climate change effects on water resources across Asian subregions.

resources across different regions of Asia (Figure 6.4), highlighting the main emerging water challenges:

- Considering the Asian region as a whole, the Tibetan Plateau (TP), often referred to as the "Water Tower of Asia," is among the largest water resources in both Asia and the world (Immerzeel et al., 2020). It plays a vital role in securing water for billions of people and supports multiple sectors, including agriculture and industry. Based on recent projections (e.g., Qiu et al., 2024), precipitation and water availability across the TP throughout the 21st century are expected to show an increasing trend in the eastern region, whereas the western part is projected to experience relatively minor changes in precipitation along with a drier trend in water availability.
- One of the most influential factors affecting agricultural water security in Central Asia is the growing water shortage driven by the retreat of mountain glaciers. This retreat leads to reduced runoff from mountain regions into downstream glacier-dependent river basins, thereby decreasing meltwater availability for surrounding agricultural areas. The Naryn River basin in Kyrgyzstan provides a clear example of this. A study by Sadyrov et al. (2025) on this basin found that, under the worst climate change scenario (SSP5-8.5) for the period 2077–2096, the overall runoff could decrease by up to 40% due to glacier retreat and increased evapotranspiration, compared to the reference period 1981–2000. Given that glaciers and snowmelt are among the most important water resources for farming activities, it is therefore crucial for the Central Asian region to adapt to these future expectations to avoid critical issues such as water and food insecurity.

- Among the regions of Asia, and indeed the world, South Asia, a significant contributor to greenhouse gas (GHG) emissions, is among the most vulnerable to climate-related changes, including rising temperatures, altered rainfall patterns, and more frequent extreme weather events such as droughts and floods (Agarwal et al., 2021). One of the most anticipated impacts in this region is major changes in the monsoon period, a key seasonal weather system that delivers huge amounts of rainfall and supports the livelihoods of billions across South Asia (Goswami et al., 1999; Wang & LinHo, 2002; Luo et al., 2024). A review by Fiaz et al. (2025) found that monsoon dynamics are shaped by complex interactions between ocean-land temperature gradients, atmospheric circulation, and topography. The study also found that while rising temperatures expected in the coming decades may increase total rainfall due to higher atmospheric moisture, a potential weakening of monsoon circulation remains a major concern.
- The most critical impacts of climate change on water resources in East Asia are largely driven by the rising frequency of extreme weather events such as floods and droughts (Mall et al., 2019). Climate change is expected to significantly threaten agricultural water security in the region due to projected shifts in precipitation patterns, including more frequent heavy rainfall events, particularly during the monsoon season, and prolonged droughts. Heavy rainfall may lead to increased soil erosion, while extended drought periods are likely to expand affected areas and reduce overall water availability. Additionally, altered precipitation regimes may decrease runoff in major catchments, further intensifying water stress across agricultural zones (Su et al., 2009). Moreover, reduced snowpack and glacier retreat in some countries of this region may reduce water flows of downstream glacier-dependent river basins needed for various uses, particularly irrigation. For instance, by 2030, compared to the year 2000, it is expected that the input of glacial meltwater to the Shule River Basin in north-western China will decrease from 23% to 15% (Wang et al., 2021). As a result, the region's agricultural systems are likely to face increasing challenges in maintaining productivity under future climate conditions.
- Southeast Asia faces growing water-related challenges due to climate change, particularly from the increasing intensity of extreme rainfall events, which are exacerbated by the limited adaptive capacity of most countries in the region (Nor Diana et al., 2022). In addition to extreme rainfall, climate change impacts in this region appear in various critical ways, including sea level rise, prolonged droughts, rising temperatures, and worsening freshwater scarcity (Hariadi et al., 2024). Looking ahead, projections by Hariadi et al. (2024) through to 2050 indicate that the region will experience greater variability in both dry and wet conditions. The study forecasts an increase in the frequency of both low-flow and high-flow events, driven by shifts in precipitation patterns. Importantly, drier conditions during low-flow periods are expected to intensify more significantly than the wetness during high-flow events, pointing to an elevated risk of agricultural water insecurity in several countries across the region.

- West Asia is expected to face a high risk of agricultural water insecurity due to the future impacts of climate change. The region, already characterized by arid and semi-arid climates, is projected to experience a warmer climate, reduced and more erratic precipitation in most countries, and increased evapotranspiration rates. Rising temperatures in this region are anticipated to alter the pattern of extreme weather events, leading to more frequent heatwaves and prolonged dry periods, even in parts where rainfall is projected to increase slightly. These changes will likely reduce overall surface and groundwater availability (Sivakumar et al., 2013). Moreover, prolonged droughts and reduced river flows, especially in transboundary basins, may lead to significant water scarcity and severely disrupt irrigation-dependent agriculture, which is dominant in many West Asian countries (e.g., Gao et al., 2021).
- In North Asia, particularly the Asian part of Russia, climate warming may bring some benefits, such as the expansion of high-quality arable land (Javeline et al., 2024). However, future climate change in this region is also expected to significantly influence agricultural water security through altered precipitation patterns, an accelerated warming trend, and the melting of snow and glaciers. These changes are likely to increase variability in water availability. As highlighted by Javeline et al. (2024), northern rivers are generally projected to experience higher flows, while southern rivers are expected to decline. Therefore, effective adaptation and water management strategies will be essential to sustain agricultural productivity under these changing conditions.

6.4.2 PROJECTIONS FOR AFRICA

Water resources in Africa are currently under serious threat due to the impacts of climate change, with serious consequences for agricultural water security, food security, and rural livelihoods (Diop et al., 2021). Looking ahead, if the continent's limited adaptive capacity to climate change persists, it is anticipated that future climate change-related conditions, such as rising temperatures, more frequent extreme weather events, and, particularly, shifts in precipitation patterns, will further intensify existing water-related challenges, including water scarcity, increased variability, reduced accessibility to water, and water quality deterioration (Diop et al., 2021).

The following points outline the potential impacts of climate change on agricultural water resources across Africa's main subregions, highlighting key emerging vulnerabilities and the urgent need for targeted adaptation strategies:

- In Southern Africa, it is anticipated that climate change will pose serious risks to agricultural water security, as it may lead to negative effects on water availability, accessibility, and demand (Kusangaya et al., 2014). Specifically, as reviewed by Kusangaya et al. (2014), it is expected that by 2050, this subregion of the African continent will experience increased temperatures (by an average of 3°C), more frequent drought and heatwave events, reduced rainfall during the growing season due to declines in soil moisture and runoff (by an average of 5–15%), and a reduction in river runoff and water availability in the dry tropics (by an average of 10–30%). These projections, combined with challenges such as a lack of skills for developing and implementing effective responses to climate change and low adaptive capacity, may lead to water and food insecurity in the region over the coming decades, unless adequate adaptation measures are implemented promptly.

• In North Africa, where agricultural water resources are already scarce, climate change is expected to further widen the gap between water supply and demand. Rising temperatures, projected to be the highest across the African continent, combined with declining rainfall (Almazroui et al., 2020), will likely increase crop water requirements while simultaneously reducing the availability of both surface and groundwater resources (Hamed et al., 2018). A study by Droogers et al. (2012), focusing on the Middle East and North Africa (MENA) region, projected that under an average climate change scenario for the period 2041–2050, total water demand will reach 393 km³ yr⁻¹, with the water shortage increasing by 157 km³ yr⁻¹. Of this shortage, 22% is attributed to climate change and 78% to socioeconomic drivers such as population growth and urbanization.

   Overall, the interplay between current climatic aridity, demographic and economic growth, and projected climate change impacts, including higher temperatures, more frequent heatwaves and droughts, and reduced precipitation, will likely lead to severe water scarcity in the coming decades. In response, local farmers may increasingly rely on groundwater abstraction to meet rising agricultural demands (Hamed et al., 2018). This situation underscores the urgent need to identify and promote the use of non-conventional water resources in North Africa, such as seawater and brackish water desalination, as well as improved rainwater harvesting techniques (e.g., El-Ghzizel et al., 2021).

• In West Africa, where agricultural water needs are primarily met by rainfall during the monsoon season (Sultan & Gaetani, 2016; Sylla et al., 2018), projected shifts in monsoon rainfall patterns and rising temperatures are expected to reduce water availability and the productivity of rainfed agriculture in some areas, particularly the western Sahel. This trend poses a serious threat to agricultural water security and local livelihoods. However, other regions, such as the eastern Sahel, are projected to become wetter (Cook et al., 2022).

   More precisely, based on UPSCALE rainfall predictions conducted by Cook et al. (2022), the Sahel is expected to become generally wetter between 2000 and 2100, with an estimated increase of about +12% in rainfall. In contrast, the western Sahel is projected to experience a decline in rainfall of about −13%. The study of Cook et al. (2022) also indicated that, due to enhanced soil hydraulic conductivity, driven by increased monsoon rainfall and reduced transpiration, groundwater recharge below 3 m depth is expected to increase. Specifically, recharge may rise from 0–16% of rainfall under current climate conditions to 1–20% under future scenarios. This improvement in groundwater recharge could enhance agricultural water security in parts of the region.

Although increased rainfall may seem promising, it is important to emphasize that it does not automatically lead to the elimination of irrigation needs or even a reduction in them across the region. Specifically, a study by Sylla et al. (2018) that examined hydrological changes in West Africa under global warming scenarios of 1.5°C and 2°C found that irrigation needs actually increased, while water availability declined across ten major river basins. This presents significant challenges for sustainable agriculture under the 2°C scenario. However, these adverse impacts were reduced by up to 50% under the 1.5°C scenario.

- In East Africa, a region rich in large water resources such as Lake Victoria, Lake Tanganyika, the Tana, Pangani, and Ruvu Rivers, as well as the Lotikipi Basin Aquifer, climate change is expected to impact agricultural water security. A study by Demissie et al. (2024) highlighted that most countries in the region, including Kenya, Uganda, and Tanzania, are projected to experience increased precipitation in the coming decades. The study also noted that extreme floods and droughts are likely to become more frequent. According to the study, higher evaporation rates could lead to more intense rainfall events and droughts, profoundly affecting the water cycle. While river flows are projected to increase in most areas, droughts may last up to 50% longer in Ethiopia, Kenya, and Tanzania, but decrease in Uganda. The study further emphasized that water deficits during droughts will worsen in Ethiopia and Kenya. Overall, these projections underscore the urgent need for effective agricultural water management to address these challenges, enhance agricultural water security, and reduce the risk of food insecurity in the region.
- Unlike Southern, Eastern, and Western Africa, Central Africa has experienced relatively consistent rainfall and river flows in recent decades (Conway et al., 2009). However, climate change is expected to result in significant rainfall variability, leading to more frequent extreme storms, floods, and droughts, particularly during the wet season, thereby posing substantial challenges for water management in agriculture (Doherty et al., 2022). Although total annual rainfall may increase, particularly in the northwest, uncertainty remains (Doherty et al., 2022). Despite these fluctuations, long-term water availability in rivers and lakes is unlikely to decline significantly by 2050 (Doherty et al., 2022). Nevertheless, due to limited investment in agricultural water storage and irrigation infrastructure, greater efforts and funding are essential to adapt to these changes and enhance both water management and agricultural productivity. Such efforts are crucial to strengthen agricultural water security and overall food security in the region.

6.4.3 PROJECTIONS FOR EUROPE

In Europe, numerous studies have projected future trends in both SW and groundwater resources, regarding both quantity and quality, under various climate change scenarios. These projections have been developed using a wide range of models and modeling approaches. The following are some examples of such studies:

- Based on the LISFLOOD hydrological model (de Roo et al., 2000), Bisselink et al. (2020) predicted that under moderate (RCP4.5) and high (RCP8.5) climate change scenarios, annual water availability is expected to decrease in Southern Europe, particularly in countries such as Spain, Portugal, Greece, Cyprus, Malta, Italy, and Turkey, with increasing trends projected for Central and Northern Europe. These findings highlight the need for effective adaptation strategies to address these changes.
- Nistor (2019) employed an analytical approach that combined effective precipitation and the De Martonne Aridity Index to assess the potential impacts of climate change on groundwater resources in Southeastern Europe. Their results revealed that around 50% of the study area is affected by drought stress, with climate change driving negative effects on groundwater resources. Furthermore, the northern, eastern, and southeastern parts of the region are projected to be the most severely affected during the period 2041–2070.
- Another study conducted by Hiscock et al. (2011), which focused on future trends in groundwater recharge, found that by 2100, enhanced recharge is expected in the northern part of Europe due to increased winter rainfall. However, this recharge will occur over a shorter time period, while summers will become drier with a longer duration of limited groundwater replenishment. This highlights the urgent need to adapt to these changes by capturing winter recharge for use during the summer months, in order to meet the increased agricultural water demand and to avoid the risk of water stress on agriculture. Conversely, reduced recharge is anticipated in the southern part of Europe, where water stress is projected to be more severe than under current climate conditions. This underscores the necessity of developing effective water management strategies to mitigate the expected negative impacts of climate change.
- Like most regions of the world, climate change is expected to alter the flow regimes of SW bodies in Europe, such as rivers and lakes. For example, using the hydrological model WaterGAP3, Schneider et al. (2013) analyzed changes in river flow regimes across Europe by comparing the periods 1971–2000 and 2041–2070. Their findings revealed that climate change is anticipated to significantly reshape natural flow regimes, particularly in the Mediterranean due to increased aridity, and in the boreal zone as a result of reduced snowmelt and rising temperatures. These projected changes may result in challenges such as seasonal water shortages and increased pressure on available water resources. In light of this, it is recommended that effective adaptation measures be developed to reduce the risk of water stress and to enhance water security in agricultural regions that depend on river water for irrigation.
- In addition to its effect on water availability, certain climate change scenarios in some European countries may lead to increased irrigation water

requirements (e.g., Riediger et al., 2014), underscoring the importance of accounting for this rise in water demand when managing irrigation resources. The most significant increases are expected in regions of Europe where the climate is projected to become drier and warmer (Döll, 2022), thereby reducing soil water availability and heightening the need for either full or supplementary irrigation.

- Not only are the flow regimes of SW bodies in Europe expected to change under climate change, but the latter may also lead to critical variations in SW quality and exacerbate the impacts of nutrient pollution on these systems (e.g., Charlton et al., 2018). In certain parts of Europe, such as the Mediterranean region, changes in rainfall patterns, increased temperatures, and decreased flows, combined with shifts in nutrient concentrations and reduced dilution capacity, are likely to further degrade SW quality, thereby diminishing its suitability for agricultural use. For instance, a study by Dorado-Guerra et al. (2023) on the Júcar River basin in Spain found that long-term increases in temperature and alterations in rainfall patterns under the climate change scenario RCP8.5 for the period 2070–2100 are projected to significantly affect SW quality in comparison to the reference period 1990–2018. More specifically, the study reported that the number of SW bodies with poor nitrate, ammonium, phosphorus, and biological oxygen demand (BOD) levels is expected to rise considerably in the future. Moreover, the researchers found that median ammonium and phosphorus concentrations could double during months characterized by low flow conditions.

Based on the above, it is clear that climate change may challenge agricultural water security across many regions of Europe, despite the overall abundance of water resources. This highlights the need to support and strengthen already implemented adaptation strategies, not only to prevent negative impacts but also to take advantage of potential opportunities.

6.4.4 PROJECTIONS FOR NORTH AMERICA

Agricultural water security under climate change in North America is expected to exhibit varying future trends, depending on the region, the climate change scenario, and the current state of water resources. These variations reflect the specific challenges faced by each area and underscore the need for the development and implementation of adaptation measures tailored to local conditions. The following points summarize some examples of these future trends and challenges:

- In the western USA, climate change is increasingly influencing the hydrology and water resources of mountainous regions. Rising temperatures are expected to intensify evapotranspiration and accelerate snowmelt, resulting in reduced snowpack and streamflow across major river basins (Das et al., 2011). Wang et al. (2025) project significant regional variation in precipitation and snowpack between 1990 and 2050: While annual precipitation is expected to increase overall, due to frequent intense precipitation events, seasonal patterns will shift, with winter precipitation increasing and summer precipitation likely to decline, especially in the High Plains. At lower elevations, snow accumulation is expected to decrease, and snowmelt may occur up to 26 days earlier. Additionally, findings by Niraula et al. (2017) suggest that groundwater recharge will increase in the northern parts of the region over the coming decades, but decline in the drier southern areas as the climate warms, which may worsen the groundwater deficit in some aquifers in these areas. Collectively, these changes highlight the substantial challenges that climate change poses for future agricultural water management and planning in the region.

- In the Great Plains (GP) and the Midwest, key agricultural regions, the challenges related to water resources differ markedly from those encountered in the western USA. In these areas, the primary concerns are associated with extreme climate events, including droughts, floods, and heatwaves. Climate model projections indicate an increase in the intensity and variability of precipitation, as well as an increased frequency of both droughts and floods (e.g., Ojima et al., 2021; Chen & Ford, 2023), with significant implications for water availability, water quality, and agricultural productivity. Consequently, there is a need to strengthen the resilience of agricultural systems in these regions through improved soil and water management strategies in response to a changing climate.

- In contrast to many subregions of North America, certain areas in Canada, such as the Prairie Provinces, may experience longer growing seasons and increased precipitation, which could potentially enhance agricultural water availability (Sauchyn et al., 2020). However, these potential benefits may be constrained by an increased risk of heatwaves, floods, droughts, and wildfires in the coming decades (Sauchyn et al., 2020), all of which could adversely impact agricultural systems. These challenges highlight the necessity of developing and implementing effective adaptation strategies to safeguard agricultural productivity under changing climatic conditions.

- In various regions of Mexico, climate change-related conditions, particularly rising temperatures and shifts in precipitation patterns, are projected to exacerbate water scarcity by increasing demand while simultaneously reducing availability, thereby resulting in critical water deficits (Hermes Ulises et al., 2022). For instance, a study by Bravo-Cadena et al. (2021), focusing on the central-eastern region of the country, projected that by 2050, under the RCP 4.5 climate change scenario, water availability could decline by approximately 15 to 26% compared to current climatic conditions. Concurrently, water demand is expected to increase, further intensifying the risk of deficits. These findings underscore the urgent need for the timely implementation of comprehensive and effective water management strategies to mitigate the anticipated impacts of future water shortages.

- Additionally, climate change may intensify competition for water among agricultural, urban, and environmental sectors, particularly in transboundary river basins such as the Colorado River and the Rio Grande. This underscores the need for greater consideration of effective cooperation in the management of shared water resources, especially under conditions of water stress (e.g., Nava & Sandoval-Solis, 2015). This is particularly critical in regions where institutional, legal, and political frameworks for water governance are insufficiently coordinated. In this context, strengthening transboundary water governance mechanisms, promoting data sharing, and fostering stakeholder engagement are essential steps towards achieving sustainable and equitable water management. Without effective and sustained cooperation, agricultural water security could be significantly compromised, potentially reaching critical levels that raise serious concerns about the availability of water for ensuring adequate and reliable food production, and, ultimately, food security for all.

6.5 LOCAL EFFECTS OF CLIMATE CHANGE

Agricultural water security is expected to show varying local trends under climate change due to the specific geographic, climatic, environmental, hydrological, and water-related conditions of each location (Sobkowiak & Wrzesiński, 2024), as well as differing socioeconomic and institutional contexts (e.g., Stefanova et al., 2019). Understanding these localized effects is essential for developing effective, context-specific adaptation measures that can simultaneously conserve water resources and enhance the resilience and adaptive capacity of agricultural systems under changing climatic conditions. In the following sections, case studies from Asia, Africa, Europe, and North America are presented to illustrate how climate change is affecting agricultural water security, such as changes in water availability, demand, and quality, under various climate change scenarios.

6.5.1 LOCAL CASE STUDIES FROM ASIA

In Asia, numerous researchers have assessed future local trends in agricultural water security under various climate change scenarios in recent years. Their studies have yielded significant insights into water availability, demand, and quality, which are critical for developing effective agricultural water management strategies. The following are findings from some of these local studies (Table 6.2).

TABLE 6.2

Examples of Studies Assessing Local Climate Change Effects on Agricultural Water Resources in Asia

Study	Study Area	Key Effect of Climate Change
Rusli et al. (2017)	Bandung groundwater basin, Indonesia	Minimal effect on groundwater recharge
Vishwakarma et al. (2024)	Bundelkhand, India	Change in crop water requirements
Song et al. (2022b)	Ganjiang River basin, China	Change in surface runoff
Nkomozepi and Chung (2014)	Geumho River basin, South Korea	Decreased water availability
Boonwichai et al. (2018)	Songkhram River basin, Thailand	Increased irrigation needs for rice

6.5.1.1   Indonesia

In addition to climate change factors such as rising temperatures and altered rainfall patterns, the manner of water use, whether sustainable or unsustainable, can exacerbate issues like groundwater recharge depletion. A study by Rusli et al. (2017), focusing on the Bandung groundwater basin in Java, Indonesia, found that climate change-related conditions, especially increased rainfall and higher temperatures, have minimal impact on groundwater recharge. By comparison, if groundwater use remains unchanged or grows with population demand, groundwater levels are projected to decline by approximately 15 m by 2050, potentially nearly doubling with increased water demand. Conversely, reducing groundwater extraction could allow levels to recover, highlighting that water use plays a more significant role in future groundwater status than climate change.

6.5.1.2   India

Climate change is expected to affect the water needs of many crops, leading to several studies across India. One of these was conducted by Vishwakarma et al. (2024), who examined future water demand in Bundelkhand, a region facing serious water problems, under climate scenarios RCP 4.5 and RCP 8.5. The study found that compared to 1982–2010, by 2071–2100 most major crops will need more water, with wheat requiring up to 27 mm more. In contrast, gram was found to be better suited for future conditions as it will not need extra irrigation. The study also showed that rainfall is likely to reduce, which could make water even scarcer. These findings highlight the importance of improved water management, especially for wheat, to prevent future farming problems. Overall, these predictions are valuable for preparing the region with the best water management and avoiding risks from poor agricultural water use.

6.5.1.3   China

Climate change-related conditions, such as rising temperatures and altered rainfall patterns, are expected to significantly affect hydrological processes, including surface runoff. As a result, many researchers in China have investigated the potential impacts of these changes. One of the most notable studies, conducted by Song et al. (2022), revealed a strong correlation between variations in climate variables and projected changes in surface runoff. Specifically, their research, carried out in the upper Ganjiang River basin, showed that a 1°C increase in temperature could reduce surface runoff depth by approximately 2.5–4.7%, while a 10% increase in precipitation could lead to a 12–14% rise in surface runoff. These projections provide valuable insights for the optimal management and planning of SW resources in the region.

6.5.1.4   South Korea

One of the primary impacts of climate change on agriculture is the change in water availability for farming activities, which directly affects agricultural productivity. In this context, a study by Nkomozepi and Chung (2014), conducted in the Geumho River basin in South Korea, examined the relative changes in mean annual runoff and the aridity index under various climate change scenarios, and found that future climatic conditions, particularly those represented by the RCP2.6, RCP4.5, and RCP8.5 scenarios, may decrease the availability of water resources in the basin compared to present conditions. This projected decrease underscores the importance of implementing more effective water conservation and management strategies to support sustainable agriculture in the face of climate change.

6.5.1.5 Thailand

Rice is one of the fundamental crops in Asia; however, its irrigation water requirements are undergoing, and are expected to continue undergoing, significant changes due to several climate-related factors. These include shifts in rainfall patterns, particularly during the reproductive stage, changes in evapotranspiration rates, alterations in rice physiology and phenology, as well as fluctuations in soil water content (Shahid, 2011). Among these factors, a study by Boonwichai et al. (2018), conducted in the Songkhram River Basin, Thailand, found that changes in rainfall patterns play a particularly critical role. Specifically, although total annual rainfall is projected to increase over the coming decades, the study showed that rainfall during the rice reproductive stage (September–October) is likely to decline under the RCP4.5 and RCP8.5 scenarios. This decrease may lead to an increased irrigation water requirement for rice. The study underscores the necessity of considering all climate change-related factors, such as rising temperatures and the timing of rainfall shifts, when assessing the impacts of climate change on crop water requirements. Such comprehensive assessments are essential for developing optimal agricultural water management strategies.

6.5.2 Local Case Studies from Africa

In Africa, numerous local studies have examined the potential impacts of climate change on agricultural water resources. Selected examples are presented below (Table 6.3).

6.5.2.1 Ethiopia

By selecting the Omo-Gibe Basin of Ethiopia as the study area and using the period 1987–2019 as a reference, Orkodjo et al. (2022) found that, due to projected increases in temperature and reductions in rainfall over the period 2010–2100, climate change scenarios RCP 4.5 and RCP 8.5 are expected to reduce annual streamflow by approximately 10%. This would lead to a decline in irrigation water availability ranging from 10% to 20%. Given the importance of this basin for agricultural production, it is essential to identify and implement effective water management strategies to address the anticipated water stress. Such measures are urgently required to prevent severe water scarcity and potential threats to food security in the region.

TABLE 6.3

Examples of Studies Assessing Local Climate Change Effects on Agricultural Water Resources in Africa

Study	Study Area	Key Effect of Climate Change
Orkodjo et al. (2022)	Omo-Gibe Basin, Ethiopia	Reduced irrigation water availability
Hirpa et al. (2018)	Turkwel River Basin, Kenya	Water scarcity
Bouras et al. (2019)	Tensift, Morocco	Changes in crop water requirements
Ochwo et al. (2025)	Kiryandongo, Uganda	Increased groundwater recharge
Obuobie et al. (2012)	Volta & Pra river basins, Ghana	Water scarcity

6.5.2.2   Kenya

Focusing on the Turkwel River Basin, one of the region's primary water sources for irrigation and other uses, Hirpa et al. (2018) found that climate change, climate variability, rapidly increasing water demand, and expanded irrigated areas could lead to significant water scarcity. Furthermore, the authors emphasized that water demand management strategies will play a pivotal role in determining the basin's future water security. These findings underscore the need for strategic and forward-looking water governance measures to mitigate the anticipated water stress and to support sustainable agricultural development in the basin.

6.5.2.3   Morrocco

As climate change is expected to influence the water requirements of many crops, a study by Bouras et al. (2019) conducted in the Tensift region of Morocco estimated that the water needs of wheat could decrease by 13–42% in the coming decades. This projected reduction is attributed to a shorter growing season driven by rising temperatures, with the extent of the decrease varying according to the climate change scenario (RCP4.5 and RCP8.5). Moreover, the study highlighted a shift in the timing of water demand, indicating that the peak water requirement is likely to occur approximately two months in advance of present-day climate conditions. These projections should therefore be taken into consideration in the irrigation management of wheat fields in the region, which may lead to more efficient agricultural water use and optimal wheat productivity.

6.5.2.4   Uganda

In Uganda, where groundwater is an essential water resource for farmers, a study by Ochwo et al. (2025), considering both climate change and land use/land cover (LULC) changes and conducted in the Kiryandongo region with the aid of the WetSpass-M model, found that groundwater recharge is projected to increase by approximately 13–14% over the period 2031–2060 under the climate change scenarios SSP1-2.6 and SSP2-4.5, mainly due to increased rainfall, relative to the baseline period 1991–2020. In contrast, when LULC remains unchanged, recharge is expected to increase by about 19 to 21%, highlighting the significant role of LULC changes in moderating future groundwater availability.

6.5.2.5   Ghana

By selecting the White Volta and Pra river basins in Ghana as the study areas, Obuobie et al. (2012) projected that, due to climate change, both basins may face critical water scarcity by 2050, potentially limiting their ability to supply sufficient water for agriculture and other activities. This underscores the urgent need to implement effective water use and management measures to cope with the anticipated future risk.

6.5.3   LOCAL CASE STUDIES FROM EUROPE

In Europe, numerous local studies have explored the potential impacts of climate change on agricultural water resources, offering insights into region-specific challenges and responses. Selected examples of these studies are presented below (Table 6.4).

TABLE 6.4

Examples of Studies Assessing Local Climate Change Effects on Agricultural Water Resources in Europe

Study	Study Area	Key Effect of Climate Change
Haidu & Nistor (2020)	Grand Est region, France	Extended high-impact groundwater areas
Sîrodoev et al. (2022)	Bălțata River basin, Moldova	Declined runoff
Riediger et al. (2014)	Uelzen, Germany	Higher irrigation needs
Lyra et al. (2024)	Almyros Basin, Greece	Intensified nitrate pollution
D'Oria et al. (2024)	Salento area, Italy	Decreased groundwater recharge

6.5.3.1 France

The negative effects of climate change-related conditions, such as rising temperatures, increased drought frequency, and changes in precipitation patterns, on groundwater resources are widely recognized on a global scale. France, like many other countries around the world, is experiencing these impacts, which are expected to intensify in the coming decades. In alignment with this concern, a study by Haidu & Nistor (2020), which focused on the Grand Est region, estimated that the areas classified as having high or very high climate-related impacts on groundwater resources will expand by the 2020s and 2050s. These findings highlight the urgent need for strengthened efforts toward effective water conservation and sustainable groundwater management.

6.5.3.2 Moldova

The changes in climate variables such as temperature and precipitation across Europe over recent decades have been substantial, and this strong trend is projected to continue in the coming years (Kovats et al., 2014), leading to significant effects on water resources, including rivers, lakes, and groundwater systems. For example, in the Bălțata River basin, Moldova, a study by Sîrodoev et al. (2022), based on SWAT simulations, projected a future decline in runoff under the RCP2.6, RCP4.5, and RCP8.5 climate scenarios, thereby reducing the availability of water resources in the region. This anticipated reduction highlights the critical need for implementing more effective water conservation and management strategies to support sustainable agriculture under changing climatic conditions.

6.5.3.3 Germany

One of the most significant aspects of climate change is the increase in temperatures, a crucial factor in crop cultivation, which alters crop water requirements and, consequently, affects the amount of irrigation water needed to apply in cultivated fields (e.g., Gabr, 2023). This effect may occur not only in arid and semi-arid regions, but also in humid areas due to specific soil and climatic conditions. For example, a study by Riediger et al. (2014), focusing on the region of Uelzen in northern Germany, where soils exhibit low water retention capacity and rainfall is sometimes inadequate, found that climate change in the region may lead to an increased need for irrigation in the future.

6.5.3.4   Greece

Europe, like many other regions of the world, is increasingly facing the dual challenge of climate change impacting both the quantity and quality of water resources. In addition to its effect on water availability, climate change is also expected to affect the quality of surface and groundwater. For example, a study by Lyra et al. (2024) on the Almyros Basin in Greece found that future climate scenarios may significantly alter the basin's water resources. Specifically, reduced precipitation, runoff, and recharge, along with rising temperatures and evapotranspiration, are projected to lower nitrogen leaching, reduce groundwater levels, and intensify nitrate pollution. These findings highlight the need for continuous groundwater monitoring and effective water management to respond to these emerging threats.

6.5.3.5   Italy

In many Mediterranean European countries, groundwater is a fundamental water resource for agricultural activities. However, it is widely recognized that climate change is already affecting, and will continue to affect, this crucial resource through changes in groundwater recharge. This phenomenon has been observed in various areas. For instance, in the Salento area of Italy, the study by D'Oria et al. (2024), based on the evaluation of meteorological indices such as the Standardized Precipitation–Evapotranspiration Index (SPEI), found that the aquifers in this region may experience decreased groundwater recharge in the future due to climate change, particularly under the RCP8.5 scenario. Therefore, it is essential to prepare for this expected impact through appropriate groundwater resource management strategies.

6.5.4   Local Case Studies from North America

In North America, a large number of local studies have investigated the possible effects of climate change on agricultural water resources, offering insights into region-specific challenges and responses. Selected examples of these studies are presented below (Table 6.5).

TABLE 6.5

Examples of Studies Assessing Local Climate Change Effects on Agricultural Water Resources in North America

Study	Study Area	Key Effect of Climate Change
Crosbie et al. (2013)	High Plains, USA	Decreased/Increased groundwater recharge
Zhao et al. (2022)	Alberta, Canada	Increased irrigation demand
Rivas et al. (2011)	Lerma-Chapala basin, Mexico	Reduced water availability
Montecelos-Zamora et al. (2018)	Cauto River, Cuba	Decreased water availability

6.5.4.1   USA

One of the most notable characteristics of climate change impacts at the local level is the potential to observe spatial differences across regions. The study by Crosbie et al. (2013) on the projected impact of climate change on groundwater recharge in the High Plains region is a good example of this. Specifically, despite uncertainties, the study showed that between 1990 and 2050, both reduced and increased recharge are expected depending on the location: A reduction of approximately 10% in the southern part and an increase of around 8% in the northern part. These findings highlight the importance of region-specific groundwater management strategies.

6.5.4.2   Canada

Even in some regions of Canada, climate change may affect irrigation water requirements, with the impacts varying by region and crop (Zhao et al., 2025). In line with this, a study by Zhao et al. (2025) focusing on the Alberta region of Canada and examining spring wheat and canola found that under climate change, particularly the SSP585 scenario, annual irrigation demand is projected to increase in the future. Additionally, the peak in monthly irrigation demand is expected to occur earlier in the season, indicating that early planting could serve as an effective adaptation strategy to mitigate this projected increase. These findings highlight the critical importance of adjusting farming schedules to better cope with the anticipated effects of climate change on water resources.

6.5.4.3   Mexico

Lakes, which are among the main SW resources supporting many human activities, are also expected to undergo significant changes in both volume and quality due to climate change. Lake Chapala, located at the downstream end of the Lerma-Chapala Basin and recognized as Mexico's largest freshwater lake, serves as a notable example. Specifically, a study by Rivas et al. (2011) projected that by 2050, surface runoff in the basin could decline by 21%, leading to a reduction in the lake's volume, which in turn may result in a deterioration of water quality.

6.5.4.4   Cuba

In Cuba, one of the few studies assessing climate change effects on surface and groundwater availability is Montecelos-Zamora et al. (2018), which focused on the Cauto River basin. Based on SWAT simulations under the RCP 8.5 scenario, the study compared the periods 1970–2000 and 2015–2039, projecting a 1.5°C temperature rise and a 38% decrease in rainfall. These changes could lead to streamflow reductions of up to 61% and aquifer recharge declines of up to 58%. These alarming projections should be seriously considered by water managers to support conservation efforts and ensure efficient water use under climate change conditions.

6.6   CLIMATE CHANGE, WATER SECURITY, AND SUSTAINABLE DEVELOPMENT GOALS

Climate change, through its growing impact on both the availability and quality of agricultural water, poses a serious threat to water security and directly hinders progress toward achieving some SDGs, such as Zero Hunger (SDG 2), Clean Water and Sanitation (SDG 6), Good Health and Well-being (SDG 3), and Life on Land (SDG 15) (Moyer & Hedden, 2020; Taka et al., 2021). Rising temperatures and increasingly erratic precipitation patterns are making agriculture, already the largest global user of freshwater, more unpredictable and vulnerable to seasonal extremes like droughts and floods. These changes can severely affect freshwater availability, reduce crop productivity, and threaten food security and rural livelihoods, particularly in regions already facing water stress. By 2050, nearly half of the global population may experience local shortages of land and water needed for food production (Ibarrola-Rivas et al., 2017). In turn, declining agricultural outputs can worsen poverty and inequality, especially for smallholder farmers who depend on rainfed farming or have limited access to irrigation infrastructure (Mashizha & Tirivangasi, 2023). Moreover, poorly managed agricultural water systems, under a changing climate, may further complicate efforts to ensure sustainable water use (Yin et al., 2025). These challenges clearly demonstrate the strong interconnection between climate change, agricultural water security, and sustainable development. Addressing them requires integrated, climate-resilient land and water management strategies that improve water efficiency and ensure equitable access. Without urgent and coordinated action, climate change will not only compromise agricultural water security but also undermine global efforts to achieve sustainable development by 2030.

6.7   CONCLUSIONS

This chapter analyzed the possible effects of climate change on agricultural water resources, revealing several important findings. First, based on case studies conducted at local, regional, and global scales reviewed here, it is evident that climate-related changes, such as rising temperatures, altered precipitation patterns, and an increase in extreme weather events, are primary drivers of the decline in agricultural water security and the growing challenges in agricultural water management across most regions worldwide. Notably, regions such as Africa and parts of Asia, which have limited adaptive capacity, may face severe water and food insecurity in the coming decades unless timely and effective adaptation strategies are implemented.

Second, the effects of climate change extend beyond changes in water quantity and quality; they may also delay progress toward achieving the United Nations SDGs, underscoring the urgent need to integrate climate resilience into development planning. This need is especially critical in regions where the impacts of climate change on agricultural water resources are expected to be more pronounced.

Finally, it is important to note that analyses of future trends in agricultural water security under climate change at various scales are subject to uncertainties. These uncertainties arise from errors in climate and hydrological models and modeling approaches, inaccuracies in observed data, calibration errors, limitations of optimization algorithms, and, sometimes, data gaps. This highlights the critical need to improve the accuracy and resolution of both climate and hydrological models to better capture complex environmental and hydrological processes, minimize errors, and provide more reliable and precise estimates of future agricultural water security under changing climatic conditions.

7 Adapting for Tomorrow
Strategies to Enhance Agricultural Water Security

7.1 INTRODUCTION

Due to a combination of climatic and non-climatic factors, agricultural water security is increasingly under threat (El Kharraz et al., 2012; Morton, 2015; Martínez-Valderrama et al., 2023). Climatic factors, particularly those associated with climate change, such as rising temperatures, prolonged droughts, more frequent floods, and altered precipitation patterns, are already impacting, and will continue to impact, agricultural water security by contributing to challenges such as increased water demand, glacier retreat, water shortages, heightened scarcity, desertification, and difficulties in water use and management (Misra, 2014). These changes affect both the quantity and quality of available freshwater resources, thereby undermining the reliability and sustainability of surface and groundwater resources essential for agricultural production and food security (Misra, 2014).

Non-climatic factors, including global population growth, globalization, urbanization, economic development, increased consumption, agricultural intensification, and land-use changes, further exacerbate the pressure on water resources (Cosgrove & Loucks, 2015). For instance, with the global population projected to reach 9.4–10.1 billion by 2050 (United Nations, 2019), water resources will face increased pressure as agricultural output must rise by up to 56% to meet the food demand of all (van Dijk et al., 2021). This situation highlights the urgent need for intensified efforts toward the conservation of available water resources to ensure sufficient water supply for agriculture and to prevent severe water management crises in the coming decades (Cosgrove & Loucks, 2015).

The aforementioned climatic and non-climatic factors not only threaten agricultural water security and disrupt crop production and food systems, but also endanger rural livelihoods, national economies, global food security, and the broader goal of sustainable development. This reality underscores the imperative for all stakeholders to manage available freshwater resources sustainably and to implement prompt and appropriate measures within agricultural systems (Russo et al., 2014; Yadav et al., 2024).

Concerns about the sustainability of water resources for agriculture in the face of those challenges underscore the urgent need for the rapid implementation of adequate adaptation strategies. Adaptation strategies tailored to specific local conditions, ranging from technical solutions, smart technologies, and efficient use of both blue and green water to institutional reforms and climate-smart agricultural policies, play a crucial role in enhancing agricultural water security (Levidow et al., 2014; Cosgrove & Loucks, 2015; Pereira et al., 2020; Seijger & Hellegers, 2023).

DOI: 10.1201/9781003660521-7

In this context, the chapter aims to explore and synthesize key adaptation strategies that can enhance future agricultural water security. Drawing on a review of current practices, emerging technologies, and policy frameworks, it identifies priority actions to support sustainable water management in agriculture. Specifically, the chapter addresses the question: What must be done to secure water for agriculture in the future? Through case studies from several countries, it highlights best practices and adaptation measures that improve water management, promote equitable distribution, and increase resilience to climate shocks, thereby contributing to food and water security for future generations.

7.2 IRRIGATION IMPROVEMENT AND STRATEGIES FOR WATER CONSERVATION

Water used for irrigation accounts for a large share of available freshwater resources, especially in arid and semi-arid regions, because irrigation is essential for food, pasture, and fiber production. Therefore, its optimal use is crucial (Koech & Langat, 2018). Specifically, approximately 70% of global freshwater is allocated to irrigate 25% of croplands, which produce 45% of the world's food (Thenkabail et al., 2011). Moreover, given that irrigated agriculture is responsible for approximately 55% of global water losses, from water sources to irrigated plots, efficient irrigation water management is urgently needed to preserve these vital resources (Zeggaf Tahiri et al., 2021). Looking ahead, rising food demand driven by population growth will increase pressure on agricultural water resources, making water-saving measures increasingly essential.

The optimal use and management of irrigation water can be achieved through the adoption of smart technologies, the implementation of efficient irrigation systems, and the reduction of water losses at all scales. Such measures can result in significant conservation of both surface and groundwater resources. In turn, this enhances the resilience of irrigated areas to water shortages and scarcity, improves the productivity of water withdrawn for irrigation, contributes to more sustainable and productive agricultural systems (Zeggaf Tahiri et al., 2021), and even alleviates the impacts of climate change and population growth (Fader et al., 2016).

There are many practices that can be applied to improve irrigation and conserve water, and each practice should be assessed and implemented carefully to achieve intended objectives such as water saving and crop yield enhancement. The following subsections present some of the most recommended practices identified by researchers (e.g., El-Nashar & Elyamany, 2023) that can significantly increase water availability, reduce irrigation requirements, increase water productivity (WP), enhance crop yields, and improve farmers' incomes.

7.2.1 CONSTRUCTION OF SURFACE AND UNDERGROUND DAMS

Dams are one of the most traditional types of hydraulic infrastructure used by humans for various purposes such as irrigation, hydropower generation, and flood control. According to the International Commission on Large Dams, there are currently more than 60,000 large dams across the globe (defined as those taller than 15 meters), with a total storage capacity of about 8767 km³, compared to less than 10,000 in 1960

(Ardila-Ardila et al., 2025). This rapid increase in the number of dams has significantly contributed to human development (Chen et al., 2016; Schmitt & Rosa, 2024). The Three Gorges Dam in China (Wilmsen & Webber, 2017), the Grand Coulee Dam in the USA (Ortolano & Cushing, 2002), and the Grand Ethiopian Renaissance Dam in Ethiopia (Abtew & Dessu, 2019) are examples of the largest dams.

From an agricultural perspective, constructing both small and large surface dams, which serve to capture and store water, offers a simple and effective strategy for addressing drought conditions, dealing with spatial and temporal variability of rainfall, and ensuring an adequate water supply for irrigated agriculture (through full irrigation) and even for rainfed agriculture (through supplemental irrigation) (Khlifi et al., 2010; Chen et al., 2016).

In addition to surface dams, in some regions, particularly arid and semi-arid ones, the construction of underground dams to store groundwater is considered a useful and cost-effective option. These structures help cope with high evaporation demand, retain rainwater within the subsurface, provide strong protection of water against evaporation and transpiration, regulate water temperature, and prevent saltwater intrusion in coastal areas (Chang et al., 2019; Baharvand et al., 2020; Rajabi et al., 2025).

Some of the main characteristics of these dams, compared to surface ones, are as follows: First, since the stored water does not submerge the surrounding area, land use near underground dams is largely unaffected. Second, they facilitate water use in areas unsuitable for surface dams due to natural constraints (e.g., unfavorable geology) or where groundwater is inaccessible (Rajabi et al., 2025).

However, it is important to note that thorough environmental, geological, and hydrogeological assessments of the construction site, as well as the development of an optimal design for underground dams (e.g., construction materials, dimensions, etc.), are essential to prevent negative impacts and guarantee their intended benefits (Chang et al., 2019; Baharvand et al., 2020; Rajabi et al., 2025). In this context, advanced technologies such as Geographic Information Systems, hydrological models, intelligent models, and economic models are highly useful, as they can process both spatial and temporal data, which must be thoroughly and carefully evaluated before initiating underground dam projects, to determine the optimal dam choice in terms of construction site, design, cost, construction time, effectiveness, and durability (Kharazi et al., 2019).

7.2.2 IRRIGATION SCHEDULING

Irrigation is applied not only in dry regions but also in rainfed croplands due to increasing variability in rainfall and temperature associated with climate change (Pereira et al., 2020). To maximize its efficiency, irrigation water should be applied to crops based on accurate scheduling, which is considered an effective approach for determining the necessary amount of water to apply at the right time (Dong, 2023). This scheduling not only helps determine the precise water needs of crops and decrease water losses during irrigation (such as through percolation), but it also enables the adjustment of irrigation frequency according to soil characteristics and the development of crop roots (Zinkernagel et al., 2020). This, in turn, helps minimize water losses, avoid issues such as soil salinity and waterlogging, improve irrigation performance, reduce energy consumption, and enhance water conservation (Song et al., 2022a; Souza & Rodrigues, 2022; El-Nashar & Elyamany, 2023).

Irrigation scheduling, i.e., the amount, duration, and frequency of irrigation, can be determined using several approaches. The most commonly used methods include soil-based approaches, which often involve monitoring the temporal variation of soil water content or soil water potential using tools such as sensors or tensiometers (Bwambale et al., 2022); weather-based approaches, where weather data (e.g., rainfall, temperature, relative humidity) are used to estimate potential evapotranspiration (ET) and crop water requirements (Pereira et al., 2020); and plant-based approaches, which rely on crop water status indicators such as canopy temperature, leaf water potential, and sap flow rates to guide irrigation application (Fernández, 2017).

Many studies have compared the aforementioned irrigation scheduling approaches under different crop, soil, and environmental conditions, assessing indicators such as water application, crop yield, and overall efficiency. For example, a study by França et al. (2024), conducted in soybean fields in São Paulo, Brazil, showed that weather- and plant-based irrigation scheduling approaches should be calibrated using soil moisture sensors to ensure optimal crop performance under varying soil and climate conditions. Specifically, the study demonstrated that when properly adjusted, these approaches can enhance grain yield and irrigation WP. Similarly, a study by Song et al. (2022a), carried out in snap bean and turfgrass fields in South Florida, USA, found that using soil moisture sensors in combination with simulation models can ameliorate irrigation efficiency by conserving water, reducing nutrient leaching into groundwater, and preserving crop yields.

Furthermore, it is worth emphasizing that recent advances in technology, such as remote sensing, artificial intelligence (AI), and machine learning, have significantly enhanced irrigation scheduling approaches. These innovations have enabled more precise, efficient, and automated water application across agricultural landscapes. Specifically, the continued evolution of remote sensing technologies, alongside improved data analysis and management capabilities, presents new possibilities for optimizing irrigation practices, particularly in contexts where water quality is not a limiting factor (Zinkernagel et al., 2020).

Moreover, the growing capabilities of Geographical Information Systems (GIS) have made it easier to integrate and analyze spatial datasets, including those related to soil properties, topography, climate, crop distribution, and water resources. As a result, GIS has become an increasingly valuable tool among irrigation researchers aiming to develop site-specific irrigation schedules. For instance, in Spain, Ramírez-Cuesta et al. (2018) employed Python programming alongside GIS to develop an advanced ArcGIS toolbox that estimates the water requirements of lettuce and peach crops. This tool combines the dual crop coefficient method with multi-source satellite imagery, providing a practical and data-driven approach to assessing crop water demand.

7.2.3 Efficient Irrigation Methods

Recently, a combination of factors, such as increasing water scarcity, the need to boost agricultural production to feed a growing population, the growing demand for more intensive agricultural practices, the limited potential for further expansion of irrigated land in most countries, and rising competition for available freshwater resources, has made the shift toward modernized and more efficient irrigation methods a paramount measure for maximizing water use efficiency (WUE) and WP. This shift contributes to the conservation and more sustainable management of freshwater resources (Evans & Sadler, 2008; Guo & Li, 2024).

Irrigation within croplands can be implemented through various methods, including flood, sprinkler, and drip irrigation. The selection of an appropriate method depends on several factors, such as crop type (e.g., vegetables or trees), irrigation requirements (high or low water needs), soil properties (e.g., texture and salinity level), land topography, climatic conditions (e.g., arid, semi-arid, or humid), water quality, and economic considerations, such as the farmer's financial capacity and the cost of irrigation infrastructure (Rogers et al., 2014; Pokhrel et al., 2018).

In recent years, owing to growing water scarcity and technological advancements, there has been a widespread shift from traditional flood irrigation to drip irrigation systems across many parts of the world (Guo & Li, 2024). Compared to other methods, drip irrigation, by delivering water uniformly and slowly to the root zone, significantly reduces water consumption (by 30–50%), minimizes soil evaporation, and enhances crop growth (Guo & Li, 2024).

7.2.4 DEFICIT IRRIGATION

Over the last few decades, deficit irrigation (DI), an approach in which irrigation is intentionally reduced below the actual crop water requirement, has been recognized as one of the most effective water-saving strategies. It is widely recognized as an effective tool for mitigating water scarcity and shortages (Levidow et al., 2014). It seeks to improve WUE by either decreasing the volume of water applied during each irrigation event or by skipping irrigation during growth stages when its impact on crop productivity is minimal (Abdelfattah & Mostafa, 2024). DI can lead to significant conservation of available freshwater resources without substantially compromising crop yields, particularly in arid and semi-arid agricultural regions where water is scarce and the effects of climate change are more pronounced (Geerts & Raes, 2009).

By applying less water than the actual crop requirement, DI also helps to improve economic WUE, minimize the severity of issues such as soil salinity and waterlogging, reduce the use of chemical inputs, and lower irrigation costs (Teshome et al., 2023). These additional benefits make DI an appealing strategy, not only for conserving water but also for enhancing the sustainability of agricultural practices in water-limited environments.

DI can be categorized into three main types: Conventional DI (CDI), where irrigation is applied at the same deficit level throughout all stages of crop growth, regulated DI (RDI), which provides full irrigation during drought-sensitive stages and reduces it during less critical periods, and partial root zone drying (PRD), where only one side of the root zone is irrigated at a time, leaving the other side dry (Khapte et al., 2019; Abdelfattah & Mostafa, 2024).

Although some studies have reported that DI may have no impact, or even negative effects, on crop yield and WP for certain crops (e.g., Ali et al., 2007; Igbadun et al., 2008; Mila et al., 2017), the majority of more recent research conducted across diverse agricultural regions worldwide supports DI as a promising strategy for enhancing water use productivity while maintaining yields and improving soil conditions, such as moisture retention, compared to full irrigation. For example, in New Deal, Texas (USA), sesame, safflower, and sunflower have been shown to perform well under DI conditions, with no significant reduction in either yield or oil content (Pabuayon et al., 2019). Similarly, in Florida (USA), Teshome et al. (2023) found that DI enabled water savings without adversely affecting the growth or productivity of sweetcorn and green beans. In Kebili, Tunisia, Haj-Amor et al. (2016) demonstrated that applying DI at 90% of crop water requirements in date palm fields improved soil moisture and reduced salinity levels, contributing to better growth and yield. Another study conducted by Patanè et al. (2011) in Sicily, Italy, revealed that, compared to full irrigation, DI of tomato at 50% of crop water requirements throughout the growing season led to greater water conservation, estimated at more than 40%, without notably affecting yield.

Several factors should be carefully evaluated prior to the implementation of DI. In particular, it is strongly recommended to identify the specific water needs of each crop and its critical growth stages, assess crop responses to DI, and evaluate whether the water availability, soil properties, and climatic conditions of the irrigated area are suitable for its successful application (Abdelfattah & Mostafa, 2024). Such assessments should be grounded in rigorous scientific research and field experimentation to ensure that DI delivers tangible benefits in terms of WP, crop yields, and soil health (Figure 7.1).

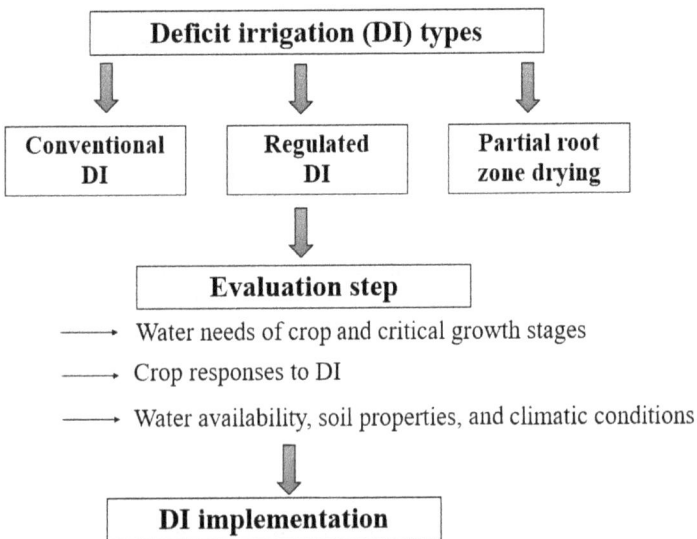

FIGURE 7.1 Main types of deficit irrigation (DI) and key factors to consider before implementation.

7.2.5 Maintenance of Irrigation Infrastructure

Regular checks, timely repairs, and proper maintenance of irrigation infrastructure components, such as dams, reservoirs, and distribution networks, are essential to detect areas of poor performance, prioritize interventions that enhance water use effectiveness, minimize water losses from the water source to irrigated plots, conserve freshwater resources, maximize the economic benefits of irrigation investments, and ensure the sustainability of irrigated farming (Bos, 1997).

In many instances, malfunctions in irrigation equipment evolve progressively, resulting in escalating repair expenditures. Therefore, the prompt identification of operational deficiencies is crucial, as it mitigates adverse effects on water availability for irrigation and agricultural productivity, while simultaneously reducing maintenance costs (Gurovich & Fernando Riveros, 2019).

Proper maintenance and modernization of irrigation equipment are particularly needed in regions where considerable water losses have been reported. For instance, in Zimbabwe, significant water losses have occurred as a result of poorly maintained irrigation infrastructure, largely due to inadequate financial and technical support, emphasizing the urgent need for state-provided assistance to enhance the performance of installed irrigation systems (Baki et al., 2025). In addition, the poor involvement of smallholder farmers in irrigation infrastructure maintenance, commonly observed in many developing countries, has contributed to the underperformance of irrigation systems, underscoring the need to address this challenge by encouraging farmer participation through training programmes, awareness campaigns, and support for local maintenance teams (Sharaunga & Mudhara, 2018).

7.2.6 Irrigation and Cropping Adaptations

In most regions worldwide, climate change-related factors, such as rising temperatures, altered precipitation patterns, and an increased frequency of droughts, have significantly influenced cropping patterns across both irrigated and rainfed agricultural systems. These impacts are reflected in shifts in the types of crops cultivated within fields, modifications in crop water requirements, changes in the length of growing seasons, and adjustments to planting and harvesting dates (Linderholm, 2006). In response, researchers around the world have proposed optimal cropping patterns, such as selecting less water-demanding crops and identifying the most suitable planting dates, to enhance irrigation efficiency, conserve water, alleviate water stress in agricultural areas, and reduce pressure on already scarce water resources. These strategies offer crops the opportunity to grow under the most favorable environmental conditions (Hamad et al., 2025).

There are several cropping pattern adjustments that can contribute to water conservation. For example, where feasible, shifting from low-value crops with high water requirements (such as cereals) to high-value crops with lower water demands (such as fruits and vegetables) can significantly reduce irrigation needs (Boser et al., 2024). In Sri Lanka, Rivera et al. (2018) found that advancing the rice growing season could lead to water savings of approximately 6%. Similarly, in the Haouz region of Morocco, Belaqziz et al. (2012) reported that early sowing of wheat could reduce irrigation water use by about 40%. Furthermore, in North China, Tang et al. (2022) demonstrated that delaying potato planting contributed to a reduction in water demand.

Taken together, these findings underscore the value of flexible and adaptive cropping practices for conserving water, particularly under current and projected climate change conditions. Promoting and supporting farmers in adopting such strategies is essential, especially given that they are low-cost and relatively easy to implement.

7.2.7 SMART IRRIGATION AND ADVANCES IN TECHNOLOGY

Recent advances in technology have significantly enhanced water management in irrigated agriculture. Aligned with the Agriculture 4.0 approach, smart irrigation practices are increasingly being implemented worldwide, supporting more efficient and sustainable water use. Through the integration of digital tools, such as smart sensors, drones, AI, and remote sensing, farmers can monitor real-time weather and soil conditions, enabling timely and precise irrigation decisions (Parra-López et al., 2025). This data-driven approach not only minimizes water losses and supports conservation but also strengthens agricultural water security. As reported in Wanyama et al. (2024), smart irrigation systems that rely on real-time environmental data can reduce water use by up to 30% compared to conventional methods. Nevertheless, adoption remains limited in certain regions, particularly in sub-Saharan Africa, where inadequate infrastructure, limited financial resources, high implementation costs, and insufficient technical support continue to pose significant barriers (Wanyama et al., 2024). Addressing these challenges is therefore essential to fully realize the benefits of smart irrigation technologies (Wanyama et al., 2024).

7.2.8 USE OF NON-CONVENTIONAL WATER RESOURCES

Due to the increasing demand for water, primarily driven by agriculture, industry, and domestic use, combined with the growing impacts of climate change on water resources, such as scarcity, shortages, and declining water quality, many farmers around the world, especially in water-scarce regions, have turned to non-conventional water (NCW) resources to irrigate certain crops. These resources include treated wastewater, desalinated water, and even saline groundwater, which are increasingly used for various agricultural activities, particularly irrigation, thereby conserving water and reducing pressure on already limited freshwater supplies (Chen et al., 2021).

The suitability of NCW for irrigating specific crops under tailored soil and water management conditions, as well as its associated environmental challenges, has been widely investigated. For example, in Tunisia, Haj-Amor et al. (2016) demonstrated that using saline water to irrigate date palms grown in sandy soils is feasible, provided that regular salt leaching is ensured through consistent irrigation at 15-day intervals throughout the year. In South Korea, Kim et al. (2020) showed that desalinated water can be safely used to irrigate lettuce without negatively affecting crop yield or causing soil salinization, as long as essential nutrients (e.g., Ca^{2+} and Mg^{2+}) are supplemented to offset losses from the desalination process and mitigate soil

TABLE 7.1

Examples of Studies Assessing the Suitability of Non-Conventional Water for Irrigation

Study	Country	Studied Crop	Used Water	Key Effect of Used Water
Haj-Amor et al. (2016)	Tunisia	Date palm	Saline water	optimized regulation of soil moisture and salinity
Kim et al. (2020)	South Korea	Lettuce	Desalinated water	Effective mitigation of soil salinity and sodicity issues
Zain Eldin et al. (2024)	Egypt	Eggplant	Magnetically treated saline water	Better control of salinity to improve crop growth

sodicity. Furthermore, in Egypt, Zain Eldin et al. (2024) found that magnetically treated saline water effectively reduces soil salinity while enhancing the growth and quality of eggplant. This is likely due to magnetic treatment altering the physical and chemical properties of water, such as reducing surface tension and modifying ion mobility (Abu-Saied et al., 2023), which improves water infiltration, nutrient uptake, and reduces salt accumulation in the root zone (Table 7.1).

Finally, it is worth noting that despite these promising results, several barriers continue to hinder the widespread and effective adoption of NCW. These include high treatment costs, reduced water quality compared to conventional resources, and, in some regions, farmers' reluctance to adopt such practices (e.g., Gómez-Ramos et al., 2024). Addressing these challenges is essential to promote the sustainable use of NCW and help safeguard freshwater resources.

7.2.9 Increased Groundwater Availability in Irrigated Areas

Due to a combination of climatic factors, particularly climate variability and climate change, and non-climatic drivers such as rapid population growth, accelerated economic development, and rising irrigation demands, many groundwater aquifers used for agriculture are facing severe depletion (Wada et al., 2010; Levintal et al., 2023). Over just 40 years, from 1960 to 2000, global groundwater depletion is estimated to have increased from approximately 126–283 km³ year⁻¹ (Wada et al., 2010). This ongoing decline has been linked to critical issues such as seawater intrusion (e.g., Alfarrah & Walraevens, 2018) and land subsidence (e.g., Hung et al., 2024), highlighting the urgent need to reduce groundwater abstraction and adopt effective recharge methods.

One of the most promising methods to enhance aquifer recharge in irrigated areas is Agricultural Managed Aquifer Recharge (Ag-MAR). This approach involves the deliberate diversion of excess surface water, such as river flows, floodwater, treated wastewater, desalinated water, or irrigation return flows, into permeable soils and aquifers during periods of water surplus, using techniques such as infiltration basins and riverbank filtration (Kourakos et al., 2019). By capturing and storing water underground, Ag-MAR not only reduces surface water losses through evaporation and runoff but also helps replenish depleted groundwater reserves, thereby improving water security for agriculture. However, it is essential that such practices are implemented in ways that minimize the risk of groundwater contamination. Long-term success requires regular monitoring of groundwater quality to ensure its suitability for intended uses and to avoid adverse impacts on human health and the environment (Levintal et al., 2023).

Seawater intrusion (SWI) is a critical process that often results in elevated salinity levels in coastal aquifers, thereby degrading water quality and diminishing both the availability and suitability of groundwater for irrigation purposes. Effective management and control of SWI in coastal irrigated areas are therefore essential. This can be achieved through an integrated approach encompassing engineering, hydrological, and policy-based interventions. Key measures include the installation of hydraulic barriers to impede the inland advancement of seawater, the regulation and reduction of excessive groundwater abstraction, and the establishment of monitoring networks to precisely track salinity levels, groundwater tables, and abstraction rates (Hussain et al., 2019).

7.3 IMPROVED WATER MANAGEMENT IN RAINFED AGRICULTURE

As rainfed agriculture is, and is expected to remain, a major contributor to global agricultural production and food security, improved water management in this sector is therefore essential to sustain and enhance its role, particularly in the face of increasing water-related risks arising from climate change, overexploitation of water resources, and inadequate water management practices. This need becomes even more critical when considering the considerable potential of this sector to enhance global food production while reducing freshwater use (Rockström et al., 2010), as well as the limited emphasis on green water within soil and water policies in regions with low irrigation application (Teferi et al., 2025).

Various practices are available to improve water management in rainfed agriculture. The following subsections present some of the most widely recommended approaches, which can substantially reduce rainfall losses, improve WP, mitigate the impacts of climate-related conditions such as droughts and floods, boost crop yields, and increase farmers' incomes.

7.3.1 RAINWATER HARVESTING

In light of the growing demand for agricultural water, particularly in rural areas, and the declining availability of freshwater resources across many regions worldwide, rainwater harvesting (RWH), has been extensively documented in the literature (e.g., Lupia et al., 2017; Sucozhañay et al., 2024) as an effective strategy for meeting the irrigation needs of urban agriculture, especially during dry periods. This approach offers a promising means of alleviating pressure on conventional rural water supplies while enhancing resilience to future water scarcity. For instance, a study by Lupia et al. (2017) demonstrated that, under high irrigation efficiency, rooftop RWH could meet the full irrigation requirements of up to 33% of urban gardens in Rome, Italy. Additionally, research by Sucozhañay et al. (2024) in Cuenca, Ecuador, highlighted the significant potential of RWH in meeting the water demands of hydroponic crops. In this context, RWH not only reduces dependence on conventional water sources but also contributes to food security in regions facing increasing soil degradation, such as erosion and salinity, since hydroponic systems are soilless cultivation methods (Sambo et al., 2019). Collectively, these findings clearly demonstrate that RWH holds substantial potential to enhance both water and food security. However, it is worth noting that the large-scale implementation of RWH systems requires a fundamental shift in Integrated Water Resources Management (IWRM), positioning rainfall as a key factor in the effective management of freshwater resources (Rockström et al., 2010). Moreover, RWH should be implemented and managed in a manner that does not reduce water availability for other sectors and areas, e.g., downstream water users (Rockström et al., 2010).

7.3.2 IMPROVED EVAPORATION MANAGEMENT

In rainfed agricultural areas, not all rainfall is utilized by crops for growth and development. A substantial portion is lost as evaporation from soil and plant surfaces, commonly referred to as non-productive evaporation. In arid and semi-arid regions, where temperatures and evaporation rates are elevated, these losses can amount to as much as 50% of total rainfall (Trisorio-Liuzzi & Hamdy, 2008). Consequently, shifting evaporative losses toward productive transpiration by crops can increase water availability in the root zone, improve green water productivity (GWP), enhance resilience to climate change-related variability, and ultimately boost agricultural output in rainfed systems across arid, semi-arid, and dry sub-humid zones (Trisorio-Liuzzi & Hamdy, 2008).

By focusing on reducing evaporation losses rather than modifying runoff patterns, improved evaporation management holds significant potential to increase productivity in rainfed agriculture while safeguarding the interests of downstream users and ecosystems. This shift from evaporation to transpiration can be achieved through several adaptation measures (Rockström et al., 2010), some of which are outlined below.

7.3.2.1 Mulching

It is a practical and easy-to-implement strategy to limit non-productive soil evaporation and conserve water. By applying a protective layer of materials, such as plastic sheets, crop residues, livestock manure, sand, or rocks, over the soil surface, mulching helps reduce soil evaporation, moisture loss, regulate soil temperature, improve soil moisture retention, and promote higher crop growth and yield. A study conducted in India by Gupta et al. (2021) reported that applying rice straw mulch to wheat fields reduced total annual soil evaporation by about 50 mm. However, before implementing this management option, the response of crops to mulching characteristics (e.g., type, amount, and thickness), as well as their effects on soil evaporation (e.g., reduction percentage or even potential increase) and rainwater infiltration, should be carefully assessed to ensure the desired outcomes and avoid negative impacts (Zribi et al., 2015).

7.3.2.2 Agroforestry

Trees in agroforestry systems (AFSs) play a key role in reducing non-productive evaporation by providing shade to cultivated crops. This shading effect leads to lower soil evaporation rates, improved soil porosity, and higher soil water content (e.g., Anderson et al., 2009; Pulido-Esquivel et al., 2025). For example, a study by Lin (2007) in Chiapas, Mexico, found that in coffee farms, a high tree canopy cover (60–80%) resulted in a 41% reduction in daily soil evaporation rates compared to farms with low canopy cover (10–30%).

7.3.2.3 Conservation Tillage

Compared to conventional tillage, conservation tillage (CT) methods, such as no-tillage, strip tillage, mulch tillage, and ridge tillage, have been widely recognized by researchers as effective tools for reducing soil disturbance, non-productive evaporation and enhancing soil physical, chemical, and biological properties, thereby contributing to greater water savings and improved crop yields (Busari et al., 2015). For example, a long-term study by Zhang et al. (2022) in spring maize fields in Shaanxi Province, China, showed that even with irregular rainfall patterns, no-tillage significantly reduced total water consumption by more than 10% and remarkably increased maize yield compared to conventional tillage. When combined with other soil management practices, such as cover crop mulching, CT can further enhance soil properties, conserve water, and boost crop yields (e.g., Niu et al., 2023).

7.3.3 Adoption of Climate-Resilient Crop Varieties

Climate change is increasing rainfall and temperature variability in most rainfed agricultural areas worldwide, a trend that is expected to continue (e.g., IPCC, 2021; Gründemann et al., 2022). Climate variability, combined with shifts in cropping patterns, means that crops in both rainfed and irrigated regions will face new environmental conditions that current varieties may not be adapted to. For example, warmer temperatures can speed up crop development, allowing the use of longer-duration, higher-yielding varieties or more intensive cropping systems (Pixley et al., 2023). To meet these challenges, it is crucial to regularly update crop varieties with traits that enhance tolerance to climate stresses and suitability for evolving cropping practices (Atlin et al., 2017; Pixley et al., 2023). Key traits include improved root development for better water uptake, increased WUE, and greater resilience to climatic extremes, factors that contribute to stabilize or increase yields and stronger performance in water-limited environments.

7.4 IMPROVED AGRICULTURAL WATER GOVERNANCE

Water governance in agriculture, encompassing the social, political, economic, and administrative dimensions of managing water resources, must be improved to effectively redefine agricultural water rights, manage available water resources equitably and efficiently, address current and future climate change impacts (e.g., water scarcity, rising demand), and ensure agricultural water security (Sismani et al., 2024).

In recent decades, numerous studies from around the world have investigated the strategies that could strengthen water governance in agriculture (e.g., Brouma & Scoullos, 2008; Hurlbert & Mussetta, 2016; Laamari et al., 2022; Luo & He, 2023; García et al., 2024; Sismani et al., 2024). Collectively, they highlight the importance of establishing robust governance structures through coordinated efforts across policy frameworks, institutional arrangements, and inclusive stakeholder participation. Specifically, they identified the following key approaches to improve the governance of agricultural water resources.

7.4.1 STRONGER INSTITUTIONAL FRAMEWORKS

Better governance starts with a clear definition of the roles and responsibilities of institutions at local, regional, and national levels. It also requires greater transparency, clear lines of accountability, active stakeholder involvement, and increased authority for local bodies and water user groups through decentralization.

7.4.2 FAIR AND EFFECTIVE AGRICULTURAL WATER PRICING

Since agriculture is the largest consumer of water in most regions and water scarcity is intensifying due to climate change and unsustainable use, all water sources, including surface water, groundwater, and treated wastewater, should be valued through effective pricing systems. These systems must be transparent, reflect the true value of water, and remain affordable, especially for small-scale and vulnerable farmers. Proper pricing encourages responsible use, reduces losses by motivating farmers to adopt water-saving practices, improves cost recovery for water services, and supports effective management of water scarcity. However, to realize these benefits, pricing policies must protect low-income groups and ecosystems by prioritizing social needs over purely economic considerations (Zetland, 2021).

7.4.3 IMPROVED LEGISLATION

Clear and enforceable laws on agricultural water rights, allocation, and use are essential at all governance levels. These laws help ensure equitable access to water, promote water savings, encourage compliance, prevent conflicts, and support responsible water use. They also protect legitimate water users from illegal extraction or misuse. Strong legal frameworks typically include clear definitions of water rights, fair and transparent allocation systems, and effective enforcement mechanisms.

7.4.4 INCREASED TRANSPARENCY AND ACCOUNTABILITY

Effective governance requires transparency and accountability at every stage of water use. This involves establishing robust monitoring systems to track consumption, distribution, and policy implementation. It also requires strong mechanisms to hold institutions and users accountable. Moreover, open access to reliable data on water availability, usage, and quality is essential to support informed decision-making and build public trust.

7.4.5 Promotion of Innovation

To enhance water conservation, decision-makers should prioritize both technical support for farmers and the promotion of innovations, whether developed by local farmers or through scientific research, focused on water-saving practices. Successful digitization of water governance and management processes depends on technical support and may involve significant costs for farmers, so these challenges must be addressed to ensure adoption (García et al., 2024).

7.4.6 Resilience to Climate Change

Given the ongoing and expected impacts of climate change on agricultural water resources, integrating climate risk into water planning and management is essential. This includes investing in adaptive infrastructure and practices, such as rainwater harvesting and efficient irrigation systems, and supporting early warning systems and climate-smart water allocation strategies to strengthen agricultural water governance.

7.5   ADDRESSING PHYSICAL AND ECONOMIC WATER SCARCITY

Effectively addressing both physical and economic water scarcity in the agricultural sector through well-designed mitigation strategies (Figure 7.2) is essential for promoting equitable, efficient, and sustainable water use among farmers, thereby strengthening both water and food security, particularly in water-scarce regions.

In addition to technical solutions, such as the promotion of efficient irrigation systems, the adoption of water-saving technologies, and improved management of soil, water, and crops under climate change, physical water scarcity, particularly in highly arid regions facing severe water stress, cannot always be addressed through conventional infrastructure such as dam construction, which may be infeasible or economically unviable. In such contexts, water scarcity can also be mitigated through support from other regions and addressed by alternative strategies and policies. These include water importation (e.g., Bazrafshan et al., 2020) and the sharing of water resources

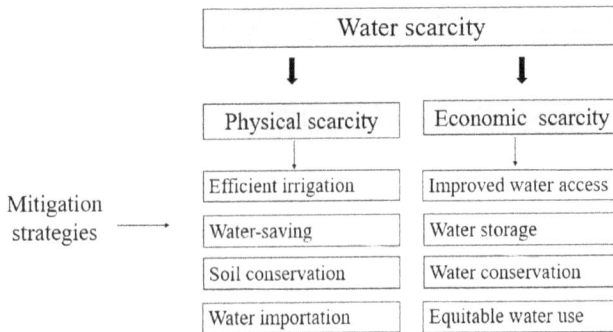

FIGURE 7.2 Examples of mitigation strategies to address physical and economic water scarcity.

through effective and cooperative water management, especially in areas that rely on shared aquifers or transboundary river basins (Degefu et al., 2016).

For water-scarce countries, agricultural policies should prioritize the export of crops with low virtual water (VW) and high economic value of water footprint (WFEV), while favoring the import of crops with high VW and low WFEV (Bazrafshan et al., 2020). This approach promotes more efficient agricultural water use and maximizes the economic returns from farming activities.

In countries that depend on shared aquifers or transboundary water resources for agriculture, cooperative water sharing mechanisms help reduce competition, prevent conflict, enhance planning for droughts, and ensure more efficient use of water, particularly during periods of scarcity. They also support the implementation of shared infrastructure and conservation strategies, contributing to long-term water security.

To cope with economic water scarcity, a multi-faceted approach is required. This includes the installation of large-scale irrigation systems to improve access to water, enhancement of water management practices, development of water storage infrastructure (e.g., small-scale water harvesting), and the promotion of water conservation. Additionally, technological advancements and policy reforms that support responsible and equitable use and distribution of agricultural water are essential (Stringer et al., 2021).

7.6 CONCLUSION

This chapter discusses the main adaptation strategies that can enhance agricultural water security in the face of emerging challenges such as climate change and population growth. Specifically, it identifies the strategies proven to be most effective in addressing agricultural water scarcity, shortages, and the increasing water demand required to meet rising food production. Despite their effectiveness, the implementation of these strategies faces significant barriers in some regions, such as inadequate hydraulic infrastructure, limited financial resources, ineffective management of some transboundary water resources, high implementation costs for certain measures, and insufficient technical support, particularly in Africa. Therefore, addressing these obstacles is essential to realize the full potential of these adaptation strategies and to secure both future water and food security. Without overcoming these challenges, progress toward sustainable agricultural water management will remain limited, threatening agricultural production, food security, and livelihoods.

Concluding Remarks and Recommendations

Water conservation in agriculture is crucial for ensuring global food security. To achieve this, it is essential first to assess the current and future availability of water resources. Based on these assessments, effective adaptation strategies must be developed, tested, validated, and implemented. These steps form the foundation for building resilient agricultural systems capable of withstanding the challenges posed by climate change and increasing water scarcity.

Building on this foundation, this book examines the current state and future prospects of agricultural water security across multiple spatial scales, from individual farms to regional, national, and global levels. Drawing on hundreds of case studies from over 70 countries, it highlights the essential role of water availability and management in sustaining agricultural production and ensuring food security. It explores how water demand, agricultural practices, and socio-economic and governance factors interact to shape water management decisions within the agriculture sector. It also analyses strategies to balance water supply and demand, improve water-use efficiency, and address the escalating challenges of population growth, climate change, and water stress, issues that significantly affect agricultural productivity worldwide.

Overall, the book presents several key findings, ranging from the climatic and non-climatic factors influencing agricultural water security, to the indicators and methodologies employed for its assessment, the current and projected levels of water availability, and the adaptation measures necessary to ensure its enhancement in the coming decades. Specifically, the concluding remarks and recommendations can be grouped into the following four thematic categories.

0.1 AGRICULTURAL WATER SECURITY ASSESSMENT

- Agricultural water security should be assessed through physical, socio-economic, governance, and political dimensions.
- Assessment should occur at multiple scales: Farm, watershed, regional, national, and global. Integrating information across these levels is crucial for developing comprehensive and coordinated action plans.
- The water footprint concept offers a useful framework for evaluating agricultural water security.
- Tools and methodologies that support public policy for sustainable agricultural water management are essential. These tools help optimize water, soil, and crop management under diverse environmental conditions. Despite their current effectiveness, improvements are needed to enhance their accuracy, adaptability, and accessibility, particularly in data-scarce regions.

DOI: 10.1201/9781003660521-8

- Evaluating the relationship between water use and availability is fundamental to ensuring future agricultural water security.
- Many agricultural water security indices overlook water quality issues, which often reduce the amount of usable water. More research is needed to incorporate water quality into water security assessments.

0.2 WATER AVAILABILITY, DEMAND, AND SCARCITY

- Water availability is a fundamental driver of agricultural production and food security. Understanding where and why water is lacking is essential for targeted interventions.
- In many regions, the balance between water demand and availability has reached critical levels, requiring urgent and sustainable water management strategies.
- Water scarcity and shortages are no longer distant threats; they are current realities that demand immediate action to secure global food production.
- Even in regions with high rainfall, agricultural water insecurity has emerged due to poor water management and governance failures.
- Limited water availability remains a key constraint to meeting the rising food demand of an increasingly affluent global population.
- Future water scarcity is expected to intensify across many regions due to rising demand, population growth, and the accelerating impacts of climate change.
- Spatial and temporal variability in freshwater availability, such as seasonal fluctuations and regional imbalances, is a major driver of water scarcity.
- Increasing dependence on groundwater has caused significant declines in groundwater storage across many regions, posing a severe risk to agricultural water security, especially in arid and semi-arid zones reliant on irrigation.

0.3 DRIVERS AND PRESSURES ON AGRICULTURAL WATER RESOURCES

- The growing global demand for food places enormous pressure on agricultural water resources.
- Agricultural water use must be optimized in response to increasing freshwater scarcity, climate change, rapid population growth, urbanization, and changing lifestyles.
- Climate change is already affecting agricultural water resources worldwide and will continue to do so, making collective efforts toward sustainable water management more urgent. Sharing local success stories can support broader global progress.
- Climate change is diminishing both the availability and quality of water for agriculture at global, regional, and local levels.
- Its impacts on water demand, availability, and quality will increasingly stress current water management systems. It is vital to assess these impacts at local and regional scales and adjust infrastructure and practices accordingly.

- Anticipated climate changes over the coming decades will likely intensify agricultural water insecurity, making robust assessments and responsive strategies essential.
- The impacts of climate change on agricultural water security will be particularly severe in developing regions, especially in parts of Africa and Asia, underscoring the need for international support to build adaptive capacity and safeguard food systems.
- In many regions, agricultural water security remains unachieved. This shortfall often exacerbates poverty and food insecurity, both of which are likely to worsen with climate change.

0.4 ADAPTATION AND MANAGEMENT STRATEGIES

- Adaptation measures must be tailored to the unique biophysical and socio-economic conditions of each agricultural zone.
- Developing effective strategies requires robust evidence on the performance of different adaptation approaches in improving water management and crop resilience.
- Sustainable management of both blue (surface and groundwater) and green (soil moisture) water is vital for food security, especially in water-scarce and vulnerable areas.
- Enhancing water use efficiency and productivity of both blue and green water is essential to feed a growing population using finite freshwater resources.
- Given the importance of rainfed agriculture in global food production, improving water management in this sector is critical. Rainfall must be treated as a key component in water planning, and failure to boost the productivity of green water may reduce yields and food availability.
- Sustainable water resource management programmes can increase agricultural productivity while reducing losses and inefficiencies.
- Efficient water use in agriculture contributes not only to food and water security but also to the achievement of various sustainable development goals (SDGs), including those related to poverty, hunger, and clean water.
- Farmers should be seen as the frontline actors in adaptation. Their decisions and practices are central to the sustainable use of agricultural water.
- In addition to technical, political, and social solutions, economic tools such as water pricing are necessary to promote the conservation of limited freshwater resources.
- Cross-sectoral cooperation is crucial to reduce competition for freshwater between agriculture, industry, and urban uses. Collaboration is also key to achieving multiple SDGs.
- Managing transboundary water resources, including rivers, lakes, and aquifers, requires attention to political and institutional factors, alongside technical and environmental considerations, to ensure equitable and sustainable use.

References

Abate BZ, AA Alaminie, TT Assefa et al. 2024. Modeling Climate Change Impacts on Blue and Green Water of the Kobo-Golina River in Data-Scarce Upper Danakil Basin, Ethiopia. J Hydrol: Reg Stud, 53:101756.

Abbas M, L Zhao, Y Wang 2022. Perspective Impact on Water Environment and Hydrological Regime Owing to Climate Change: A Review. Hydrology, 9(11):203.

Abbas SA, RT Bailey, JT White et al. 2024. A Framework for Parameter Estimation, Sensitivity Analysis, and Uncertainty Analysis for Holistic Hydrologic Modeling Using SWAT+. Hydrol Earth Syst Sci, 28:21–48.

Abbott BW, K Bishop, JP Zarnetske et al. 2019. Human Domination of the Global Water Cycle Absent from Depictions and Perceptions. Nat Geosci, 12:533–540.

Abdelfattah A, H Mostafa 2024. Potential of Soil Conditioners to Mitigate Deficit Irrigation Impacts on Agricultural Crops: A Review. Water Resour Manage, 38:2961–2976.

Abraham T, A Muluneh 2022. Quantifying Impacts of Future Climate on the Crop Water Requirement, Growth Period, and Drought on the Agricultural Watershed, in Ethiopia. Air, Soil and Water Research, 15(4):1–15.

Abtew W, SB Dessu 2019. The Grand Ethiopian Renaissance Dam on the Blue Nile. Springer International Publishing: Gewerbesraße. https://doi.org/10.1007/978-3-319-97094-3

Abulude I, S Wahlen 2024. Food Loss Analysis in Nigeria: A Systematic Literature Review. Environ Chall, 17:101027.

Abu-Saied MA, EA El Desouky, ME Abou Kamer et al. 2023. Influence of Magnetic Field on the Physicochemical Properties of Water Molecule Under Growing of Cucumber Plant in an Arid Region. J King Saud Univ, 35:102890.

Acharki S, S Taia, Y Arjdal et al. 2023. Hydrological Modeling of Spatial and Temporal Variations in Streamflow Due to Multiple Climate Change Scenarios in Northwestern Morocco. Clim Serv, 30:100388.

Adane ZA, JB Gates 2015. Determining the Impacts of Experimental Forest Plantation on Groundwater Recharge in the Nebraska Sand Hills (USA) Using Chloride and Sulfate. Hydrogeol J, 23:81–94.

Adejuwon JO, E Dada 2021. Temporal Analysis of Drought Characteristics in the Tropical Semi-Arid Zone of Nigeria. Sci Afr, 14:e01016.

Adom RK, MD Simatele 2024. Overcoming Systemic and Institutional Challenges in Policy Implementation in South Africa's Water Sector. Sustain Water Resour Manag, 10:69.

Agarwal R, V Balasundharam, P Blagrave et al. 2021. "Climate Change in South Asia: Further Need for Mitigation and Adaptation," IMF Working Papers 2021/217, International Monetary Fund.

AghaKouchak A, E Habib 2010. Application of a Conceptual Hydrologic Model in Teaching Hydrologic Processes. Int J Engng Ed, 26(4):963–973.

Agrawal M, SS Deepak 2000. Elevated Atmospheric Carbon Dioxide and Plant Responses. In: Yunus, M, N Singh, LJ de Kok (eds), Environmental Stress: Indication, Mitigation and Eco-Conservation. Springer, Dordrecht. https://doi.org/10.1007/978-94-015-9532-2_8.

Ahammed SJ, ES Chung, S Shahid 2018. Parametric Assessment of Pre-Monsoon Agricultural Water Scarcity in Bangladesh. Sustainability, 10(3):819.

Ahmad AY, MA Al-Ghouti 2020. Approaches to Achieve Sustainable Use and Management of Groundwater Resources in Qatar: A Review. Groundw Sustain Dev, 11:100367.

Ahmed N, J Hornbuckle, GM Turchini 2022. Blue–Green Water Utilization in Rice–Fish Cultivation Towards Sustainable Food Production. Ambio, 51:1933–1948. https://doi.org/10.1007/s13280-022-01711-5

Ahmed S. 2018. Analysis of Daily Precipitation Data from Selected Sites in the United States. Theses, Dissertations and Culminating Projects, 116. https://digitalcommons.montclair.edu/etd/116

Ajjur SB, SG Al-Ghamdi 2022. Towards Sustainable Energy, Water and Food Security in Qatar Under Climate Change and Anthropogenic Stresses. Energy Rep, 8:514–518.

Alejo LA, AS Alejandro 2022. Changes in Irrigation Planning and Development Parameters Due to Change. Water Resour Manage, 36:1711–1726.

Alfarrah N, K Walraevens 2018. Groundwater Overexploitation and Seawater Intrusion in Coastal Areas of Arid and Semi-Arid Regions. Water, 10(2):143.

Al-Huwaishel AS, A Elmi, A Mukhopadhyay 2022. Aquifer Storage of Treated Wastewater for Subsequent Recovery as an Important Strategy for Sustainable Water Security in Kuwait. Water Supply, 22(2):2067–2081.

Ali MH, MR Hoque, AA Hassan et al. 2007. Effects of Deficit Irrigation on Yield, Water Productivity, and Economic Returns of Wheat. Agric Water Manag, 92:151–161.

Aligholi F, D Hayati 2022. Agricultural Water Security from the Perspective of Critical Theory Paradigm. Front Water, 4:964688. https://doi.org/10.3389/frwa.2022.964688.

Al-Jayyousi OR 2003. Greywater Reuse, Towards Sustainable Water Management. Desalination, 156(1):181–192.

Allan JA. 2010. Prioritizing the Processes Beyond the Water Sector That Will Secure Water for Society—farmers, Fair International Trade and Food Consumption and Waste. In: Martinez-Cortina L, A Garrido, E Lopez-Gunn (eds), Rethink Water Food Secur. Fourth Botín Foundation Water Workshop (pp. 93–106). CRC Press, Leyden, Netherland.

Allan RP, E Hawkins, N Bellouin et al. 2021. IPCC: Summary for Policymakers. In: Masson-Delmotte V, P Zhai, A Pirani et al. (eds), Climate Change 2021: The Physical Science Basis. Working Group I Contribution to the IPCC Sixth Assessment Report. Cambridge University Press, Cambridge, UK.

Allan T. 2001. The Middle East Water Question: Hydropolitics and the Global Economy. I.B. Tauris & Co, Ltd, London, UK; New York, NY, USA.

Allen LH Jr, VG Kakani, JC Vu et al. 2011. Elevated CO_2 Increases Water Use Efficiency by Sustaining Photosynthesis of Water-Limited Maize and Sorghum. J Plant Physiol, 168(16):1909–1918.

Almasalmeh O, AA Saleh, KA Mourad 2022. Soil Erosion and Sediment Transport Modelling Using Hydrological Models and Remote Sensing Techniques in Wadi Billi, Egypt. Model Earth Syst Environ, 8:1215–1226.

Almazroui M, F Saeed, S Saeed et al. 2020. Projected Change in Temperature and Precipitation Over Africa from CMIP6. Earth Syst Environ, 4:455–475.

Aloui S, A Zghibi, A Mazzoni et al. 2023. Groundwater Resources in Qatar: A Comprehensive Review and Informative Recommendations for Research, Governance, and Management in Support of Sustainability. J Hydrol: Reg Stud, 50:101564.

Amanullah S, Khalid, Imran et al. 2020. Effects of Climate Change on Irrigation Water Quality. In: Fahad, S, M Hasanuzzaman, M Alam (eds), Environment, Climate, Plant and Vegetation Growth (pp. 123–132). Springer International Publishing, Cham.

Amirabadizadeh M, AH Ghazali, YF Huang et al. 2017. Assessment of Impacts of Future Climate Change on Water Resources of the Hulu Langat Basin Using the Swat Model. Water Harvesting Research, 2(2):13–29.

Amparo-Salcedo M, A Pérez-Gimeno, J Navarro-Pedreño 2025. Water Security Under Climate Change: Challenges and Solutions Across 43 Countries. Water, 17(5):633.

Anderson SH, RP Udawatta, T Seobi et al. 2009. Soil Water Content and Infiltration in Agroforestry Buffer Strips. Agrofor Syst, 75:5–16.

Ardila-Ardila YV, ID Gómez-Araújo, JD Villalba-Morales et al. 2025. Effect of Environmental Factors on Modal Identification of a Hydroelectric Dam's Hollow-Gravity Concrete Block. J Civil Struct Health Monit, 15:777–794.

Arnold JG, R Srinivasan, RS Muttiah et al. 1998. Large Area Hydrologic Modeling and Assessment Part 1: Model Development. J Am Water Resour. Assoc, 34:73–89.

Arreguín-Cortéz F, M Lopez-Perez, H Marengo-Mogollon 2011. Water Resources in Mexico, Water Resources in Mexico: Scarcity, Degradation, Stress, Conflicts, Management, and Policy, Hexagon Series on Human and Environmental Security and Peace. Springer: Berlin. https://doi.org/10.1007/978-3-642-05432-7.

Aryal JP, TB Sapkota, R Khurana et al. 2020. Climate Change and Agriculture in South Asia: Adaptation Options in Smallholder Production Systems. Environ Dev Sustain, 22:5045–5075.

Ashraf S, A Nazemi, A AghaKouchak 2021. Anthropogenic Drought Dominates Groundwater Depletion in Iran. Sci Rep, 11:9135.

Ashwin KRN, S Arulmozhi, A Gopalan et al. 2022. Correlation, Regression Analysis, and Spatial Distribution Mapping of WQI for an Urban Lake in Noyyal River Basin in the Textile Capital of India. Adv Mater Sci Eng, 2022:3402951.

Asresu AT, E Furlan, F Horneman et al. 2025. A Systematic Review of Climate Change Impacts on Water Quality in Transitional Environments from a Multi-Hazard Perspective. Estuar Coast Shelf Sci, 317:109194.

Atlin GN, JE Cairns, B Das 2017. Rapid Breeding and Varietal Replacement are Critical to Adaptation of Cropping Systems in the Developing World to Climate Change. Glob Food Sec, 12:31–37.

Avazdahandeh S, S Khalilian 2021. The Effect of Urbanization on Agricultural Water Consumption and Production: The Extended Positive Mathematical Programming Approach. Environ Geochem Health, 43(1):247–258.

Aznarez C, P Jimeno-Sáez, A López-Ballesteros et al. 2021. Analysing the Impact of Climate Change on Hydrological Ecosystem Services in Laguna del Sauce (Uruguay) Using the SWAT Model and Remote Sensing Data. Remote Sensing, 13(10):2014.

Baalousha HM 2016. Using Monte Carlo Simulation to Estimate Natural Groundwater Recharge in Qatar. Model Earth Syst Environ, 2:87.

Baggio G, M Qadir, V Smakhtin 2021. Freshwater Availability Status Across Countries for Human and Ecosystem Needs. Sci Total Environ, 792:148230.

Bagley JE, RD Ankur, JH Keith et al. 2014. Drought and Deforestation: Has Land Cover Change Influenced Recent Precipitation Extremes in the Amazon? J Climate, 27:345–361.

Baharvand S, J Rahnamarad, S Soori 2020. Assessment of the Potential Areas for Underground Dam Construction in Roomeshgan, Lorestan Province. Iran J Earth Sci, 12(1):32–41.

Bahramifard A, M Zibaei 2024. Integrated Assessment of Water Security in D-8 Countries. Heliyon, 10(21):e39781.

Baki CB, A Keïta, S Palé et al. 2025. Community Management of Irrigation Infrastructure in Burkina Faso: A Diagnostic Study of Six Dam-Adjacent Irrigation Areas. Agriculture, 15(5):477.

Balasubramanya S, D Stifel 2020. Viewpoint: Water, Agriculture & Poverty in an Era of Climate Change: Why do We Know so Little? Food Policy, 93:101905.

Baran-Gurgul K, A Rutkowska 2024. Water Resource Management: Hydrological Modelling, Hydrological Cycles, and Hydrological Prediction. Water, 16(24):3689.

Barbetta S, B Bonaccorsi, S Tsitsifli et al. 2022. Assessment of Flooding Impact on Water Supply Systems: A Comprehensive Approach Based on DSS. Water Resources Management, 36:5443–5459. https://doi.org/10.4060/cc7900en

Barresi Armoa OL, JG Arnold, K Bieger et al. 2024. Large Wetlands Representation in SWAT+: The Case of the Pantanal in the Paraguay River Basin. Front Water, 6:1451648.

Bateki CA, SE Wassie, A Wilkes 2023. The Contribution of Livestock to Climate Change Mitigation: A Perspective From a Low-Income Country. Carbon Manage, 14(1):1–16.

Bazrafshan O, H Zamani, HR Etedali et al. 2020. Improving Water Management in Date Palms Using Economic Value of Water Footprint and Virtual Water Trade Concepts in Iran. Agric Water Manag, 229:105941.

Belaqziz S, S Khabba, MH Kharrou et al. 2012. Optimizing the Sowing Date to Improve Water Management and Wheat Yield in a Large Irrigation Scheme, through a Remote Sensing and an Evolution Strategy-Based Approach. Remote Sensing, 13(18):3789.

Benaafi M, A Pradipta, B Tawabini et al. 2024. Suitability of Treated Wastewater for Irrigation and Its Impact on Groundwater Resources in Arid Coastal Regions: Insights for Water Resources Sustainability. Heliyon, 10(8):e29320.

Benabderrazik K, B Kopainsky, L Tazi et al. 2021. Agricultural Intensification can no Longer Ignore Water Conservation–A Systemic Modelling Approach to the Case of Tomato Producers in Morocco. Agric. Water Manag, 256:107082.

Bergstrom S 1995. The HBV Model. In: Singh, VP (ed), Computer Models of Watershed Hydrology (pp. 443–476). Water Resources Publications, Highlands Ranch.

Bett B, J Lindahl, G Delia 2019. Climate Change and Infectious Livestock Diseases: the Case of Rift Valley Fever and Tick-Borne Diseases. In: Rosenstock, T, A Nowak, E Girvetz (eds), The Climate-Smart Agriculture Papers. Springer, Cham. https://doi.org/10.1007/978-3-319-92798-5_3

Bharathi P, R Dayana, B Sivani. 2023. Valorization of Food Waste into Biofertilizer and Enhancement of Anaerobic Digestion Process Using Nanocatalyst. Biomass Conv Bioref. https://doi.org/10.1007/s13399-023-05062-3

Bhatt R, SS Kukal, MA Busari et al. 2016. Sustainability Issues on Rice-Wheat Cropping System. Int Soil Water Conserv Res, 4:68–83.

Biazin B, G Sterk, M Temesgen et al. 2012. Rainwater Harvesting and Management in Rainfed Agricultural Systems in Sub-Saharan Africa – A Review. Physics and Chemistry of the Earth, Parts A/B/C, 47:139–151.

Biondi D, G Freni, V Iacobellis et al. 2012. Validation of Hydrological Models: Conceptual Basis, Methodological Approaches and a Proposal for a Code of Practice. Physics and Chemistry of the Earth, Parts A/B/C, 42-44:70–76.

Bisselink B, J Bernhard, E Gelati et al. 2020. Climate Change and Europe's Water Resources, EUR 29951 EN, Publications Office of the European Union, Luxembourg. https://doi.org/10.2760/15553, JRC118586

Biswas A, S Sarkar, S Das et al. 2025. Water Scarcity: A Global Hindrance to Sustainable Development and Agricultural Production – A Critical Review of the Impacts and Adaptation Strategies. Cambridge Prisms: Water, 3:e4.

Biswas, AK 2003. Water and Agriculture. In: Biswas, AK (ed), Water Resources of North America. Springer, Berlin, Heidelberg. https://doi.org/10.1007/978-3-662-10868-0_6

Blagrave K, L Moslenko, UT Khan 2022. Heatwaves and Storms Contribute to Degraded Water Quality Conditions in the Nearshore of Lake Ontario. J Great Lakes Res, 48:903–913.

Blanc, E, I Noy 2023. Impacts of Droughts and Floods on Agricultural Productivity in New Zealand as Measured from Space. Environ Res: Climate, 2:035001.

Blanco-Gómez P, P Jimeno-Sáez, J Senent-Aparicio et al. 2019. Impact of Climate Change on Water Balance Components and Droughts in the Guajoyo River Basin (El Salvador). Water, 11(11):2360.

Boone RB, RT Conant, J Sircely et al. 2018. Climate Change Impacts on Selected Global Rangeland Ecosystem Services. Glob Change Biol, 24:1382–1393.

Boonwichai S, S Shrestha, MS Babel et al. 2018. Climate Change Impacts on Irrigation Water Requirement, Crop Water Productivity and Rice Yield in The Songkhram River Basin, Thailand. J Cleaner Prod, 198:1157–1164.

Borrelli P, DA Robinson, P Panagos et al. 2020. Land use and Climate Change Impacts on Global Soil Erosion by Water (2015–2070). Proc Natl Acad Sci U S.A., 117(36): 21994–22001. https://doi.org/10.1073/pnas.2001403117

Bos MG 1997. Performance Indicators for Irrigation and Drainage. Irrig Drain Syst, 11:119–137.

Boser A, K Caylor, A Larsen et al. 2024. Field-Scale Crop Water Consumption Estimates Reveal Potential Water Savings in California Agriculture. Nat Commun, 15(1):2366.

Boughton WC 1984. A Simple Model for Estimating the Water Yield of Ungauged Catchments. Civ Eng Trans Inst Eng Canberra, CE26(2):83–88.

Bouras E, L Jarlan, S Khabba et al. 2019. Assessing the Impact of Global Climate Changes on Irrigated Wheat Yields and Water Requirements in a Semi-Arid Environment of Morocco. Sci Rep, 9:19142.

Bravo-Cadena J, NP Pavón, P Balvanera et al. 2021. Water Availability–Demand Balance under Climate Change Scenarios in an Overpopulated Region of Mexico. Int J Environ Res Pub Health, 18(4):1846.

Broberg MC, P Högy, H Pleijel 2017. CO_2-Induced Changes in Wheat Grain Composition: Meta-Analysis and Response Functions. Agronomy, 7(2):32.

Brouma AD, MJ Scoullos 2008. Water Governance in the Mediterranean Region and Public Involvement. Water Mediterr. 2008:122–132.

Budhathoki BR, TR Adhikari, S Shrestha et al. 2023. Application of hydrological model to simulate streamflow contribution on water balance in Himalaya river basin, Nepal. Front. Earth Sci, 11:1128959.

Burnash RJE, RL Ferral, RA McGuire 1973. A Generalized Streamflow Simulation System. Joint Federal State River Forecast Centre, Sacramento.

Burt CM, AJ Clemmens, TS Strelkoff et al. 1997. Irrigation Performance Measures: Efficiency and Uniformity. J Irrig. Drain Eng, 123:423–442.

Burt TP, NJK Howden, F Worrall 2014. On the Importance Of Very Long-Term Water Quality Records. Wiley Interdiscip. Rev. Water, 1:41–48.

Busari MA, SS Kukal, A Kaur et al. 2015. Conservation Tillage Impacts On Soil, Crop And The Environment. Int. Soil Water Conserv. Res, 3(2):119–29.

Bwambale E, FK Abagale, GK Anornu 2022. Smart Irrigation Monitoring And Control Strategies For Improving Water Use Efficiency In Precision Agriculture: A Review. Agric. Water Manag, 260:107324.

Byaruhanga N, D Kibirige, S Gokool et al. 2024. Evolution of Flood Prediction and Forecasting Models for Flood Early Warning Systems: A Scoping Review. Water, 16:1763.

Cai X, X Zhang, PH Noël et al. 2015. Impacts of Climate Change On Agricultural Water Management: A Review. WIRES Water, 2(5):439–455.

CARD 2019. SWAT Literature Database for Peer-Reviewed Journal Articles; Center for Agricultural and Rural Development. Iowa State University, Ames, IA, USA.

Castellazzi P, D Burgess, A Rivera et al. 2019. Glacial Melt and Potential Impacts On Water Resources In The Canadian Rocky Mountains. Water Resour. Res, 55:10191–10217.

Chagas VBP, PLB Chaffe, G Blöschl 2022. Climate and Land Management Accelerate The Brazilian Water Cycle. Nat Commun, 13(1):5136.

Chang Q, T Zheng, X Zheng et al. 2019. Effect of Subsurface Dams on Saltwater Intrusion And Fresh Groundwater Discharge. J. Hydrol, 576:508–519.

Charlton MB, MJ Bowes, MG Hutchins et al. 2018. Mapping Eutrophication Risk from Climate Change: Future Phosphorus Concentrations In English Rivers. Sci. Total Environ, 613–614:1510–1526.

Chen CY, SW Wang, H Kim et al. 2021. Non-Conventional Water Reuse in Agriculture: A Circular Water Economy. Water Res, 199:117103.

Chen J, H Shi, D Sivakumar et al. 2016. Population, Water, Food, Energy and Dams. Renew. Sust. Energ. Rev, 56:18–28.

Chen L, TW Ford 2023. Future Changes in the Transitions of Monthly-To-Seasonal Precipitation Extremes Over the Midwest in Coupled Model Intercomparison Project Phase 6 Models. Int. J. Climatol, 43(1):255–274.

Chidiac S, P El Najjar, N Ouaini et al. 2023. A Comprehensive Review of Water Quality Indices (Wqis): History, Models, Attempts and Perspectives. Rev. Environ. Sci. Biotechnol, 22(2):349–395.

Chiew F, T McMahon 1994. Application of the Daily Rainfall Runoff Model Modhydrolog To 28 Australian Catchments. J. Hydrol, 153:383–416.

Ching YC, YH Lee, ME Toriman et al. 2015. Effect of The Big Flood Events on the Water Quality of The Muar River, Malaysia. Sustain. Water Resour. Manag, 1:97–110.

Chouchane H, MS Krol, AY Hoekstra 2020. Changing Global Cropping Patterns to Minimize National Blue Water Scarcity. Hydrol. Earth Syst. Sci, 24:3015–3031.

Chowdhuri I, SC Pal, A Saha et al. 2022. Field-Based Index of Land Suitability (Ils): A New Approach For Rainfed Paddy Crop Production In Groundwater Scarce Region. Geocarto. Int, 37(27):16803–16826.

Colín-García G, E Palacios-Vélez, A López-Pérez et al. 2024. Evaluation of the Impact of Climate Change on the Water Balance of the Mixteco River Basin with the SWAT Model. Hydrology, 11(4):45.

Collins DBG 2020. New Zealand River Hydrology under Late 21st Century Climate Change. Water, 12(8):2175.

Connolly-Boutin L, B Smit 2016. Climate Change, Food Security, And Livelihoods in Sub-Saharan Africa. Reg. Environ. Change, 16:385–399. https://doi.org/10.1007/s10113-015-0761-x.

Conway D, A Persechino, S Ardoin-Bardin et al. 2009. Rainfall and Water Resources Variability in Sub-Saharan Africa during the Twentieth Century. J. Hydrometeorol, 10:41–59.

Cook PA, ECL Black, A Verhoef, et al. 2022. Projected Increases in Potential Groundwater Recharge and Reduced Evapotranspiration Under Future Climate Conditions in West Africa. J. Hydrol. Reg. Stud, 41:101076.

Cosgrove WJ, DP Loucks 2015. Water Management: Current And Future Challenges and Research Directions. Water Resour. Res, 51(6):4823–4825.

Cosgrove WJ, DP Loucks 2015. Water management: Current and Future Challenges and Research Directions. Water Resour. Res, 51(6):4823–39.

Costabile P, F Macchione 2015. Enhancing River Model Set-Up For 2-D Dynamic Flood Modelling. Environ. Model Softw, 67:89–107.

Coudard A, E Corbin, J de Koning et al. 2021. Global Water and Energy Losses from Consumer Avoidable Food Waste. J. Clean. Prod, 326:129342.

Crawford NH, RK Linsley 1966. Digital Simulation in Hydrology: Stanford Watershed Model IV, Report 39, Department of Civil Engineering, CA, USA.

Crosbie RS, BR Scanlon, FS Mpelasoka et al. 2013. Potential Climate Change Effects On Groundwater Recharge In The High Plains Aquifer, USA. Water Resour. Res, 49(7):3936–3951.

Csáki P, K Gyimóthy, P Kalicz et al. 2020. Multi-Model Climatic Water Balance Prediction in The Zala River Basin (Hungary) Based On A Modified Budyko Framework. J. Hydrol. Hydromech, 68:200–210.

D'Ambrosio E, F Gentile, AM De Girolamo 2019. Assessing the Sustainability in Water Use At The Basin Scale Through Water Footprint Indicators. J. Clean Prod, 118847.

D'Odorico P, DD Chiarelli, L Rosa et al. 2020. The Global Value of Water in Agriculture. Proc. Natl. Acad. Sci. USA, 117(36):21985–21993.

D'Oria M, G Balacco, V Todaro et al. 2024. Assessing the Impact of Climate Change on a Coastal Karst Aquifer in a Semi-Arid Area. Groundw. Sustain. Dev, 25:101131.

Dallison RJH, AP Williams, IM Harris et al. 2022. Modelling the Impact of Future Climate Change on Streamflow and Water Quality in Wales, UK. Hydrol. Sci. J, 67(6):939–962.

Dang NM, VT Tu, M Babel, et al. 2021. Water Security Assessment for the Red River Basin, Vietnam. In: Ribbe, L., Haarstrick, A., Babel, M., Dehnavi, S., Biesalski, H.K. (eds), Towards Water Secure Societies. Springer, Cham. https://doi.org/10.1007/978-3-030-50653-7_2.

Dao PU, AG Heuzard, TXH Le, et al. 2024. The Impacts of Climate Change on Groundwater Quality: A Review. Sci. Total Environ. 912, 169241.

Das T, D Pierce, D Cayan et al. 2011. The Importance of Warm Season Warming to Western Us Streamflow Changes. Geophys. Res. Lett, 2011:L23403.

Davis KF, MC Rulli, A Seveso et al. 2017. Increased Food Production and Reduced Water Use Through Optimized Crop Distribution. Nature Geosci, 10:919–924.

De Bhowmick G, M Hayes 2023. Potential Of Seaweeds to Mitigate Production of Greenhouse Gases During Production Of Ruminant Proteins. Glob. Chall, 7(5):2200145. https://doi.org/10.1002/gch2.202200145.

Deen TA, MA Arain, O Champagne et al. 2025. Blue And Green Water Scarcity in the Mckenzie Creek Watershed of the Great Lakes Basin. Hydrol. Process, 39:e70038.

Degefu DM, W He, L Yuan et al. 2016. Water Allocation in Transboundary River Basins under Water Scarcity: A Cooperative Bargaining Approach. Water Resour. Manage, 30:4451–4466.

de Jager A, C Corbane, F Szabo 2022. Recent Developments in Some Long-Term Drought Drivers. Climate, 10(3):31.

Demissie T, S Gebrechorkos, M Radeny et al. 2024. Climate Risk Profile for East Africa. International Livestock Research Institute (ILRI), Nairobi, Kenya.

Deng Z, Y Hu, X Wang et al. 2025. Transitioning to Healthy and Sustainable Diets Has Higher Environmental and Affordability Trade-Offs for Emerging and Developing Economies. Nat. Commun, 16:3948.

de Oliveira Veras M, E Parenti, SS Neiva et al. 2021. Food Security: Conceptual History and Pillars. In: Leal Filho, W, AM Azul, L Brandli (eds), Zero Hunger. Encyclopedia of the UN Sustainable Development Goals. Springer, Cham. https://doi.org/10.1007/978-3-319-69626-3_21-1.

de Roo APJ, CG Wesseling, WPA van Deursen 2000. Physically Based River Basin Modelling within a GIS: The LISFLOOD Model. Hydrol. Process, 14:1981–1992.

Devak M, C Dhanya 2017. Sensitivity Analysis of Hydrological Models: Review and Way Forward. J. Water Clim, 8:557–575.

Devia GK, BP Ganasri, GS Dwarakish 2015. A Review on Hydrological Models. Aquatic Procedia, 4:1001–1007.

Dhawale R, CJ Schuster-Wallace, A Pietroniro 2024. Assessing the Multidimensional Nature of Flood and Drought Vulnerability Index: A Systematic Review of Literature. Int. J. Disaster Risk Reduct, 112:104764.

DHI. 2005. MIKE SHE Water Movement. User Guide and Technical Reference Manual, edition 1.1. Danish Hydraulic Institute. http://www.dhi.dk

Dieter CA, MA Maupin, RR Caldwell et al. 2018. Estimated use of water in the United States in 2015. U.S. Geological Survey Circular 1441, 65 p. https://doi.org/10.3133/cir1441.

Dinar A, A Tieu, H Huynh 2019. Water Scarcity Impacts on Global Food Production. Glob Food Secur, 23:212–226. https://doi.org/10.1016/j.gfs.2019.07.007.

Diop S, P Scheren, A Niang 2021. Climate Change and Water Resources in Africa. Springer Cham. https://doi.org/10.1007/978-3-030-61225-2.

Ditthakit P, S Pinthong, N Salaeh et al. 2021. Using Machine Learning Methods for Supporting GR2M Model in Runoff Estimation in an Ungauged Basin. Sci. Rep, 11:19955.

Dixon SJ, DA Scar, NA Odoni et al. 2016. The Effects of River Restoration on Catchment Scale Flood Risk and Flood Hydrology. Earth Surf. Proc. Land, 41:997–1008.

Doherty A, P Megan, C Roger et al. 2022. Climate Risk Report for the Central Africa Region. Met Office, ODI, FCDO.

Döll P 2009. Vulnerability to the Impact of Climate Change on Renewable Groundwater Resources: A Global-Scale Assessment. Environ. Res. Lett, 4:035006.

Döll P 2002. Impact of Climate Change and Variability on Irrigation Requirements: A Global Perspective. Clim. Chang, 54:269–293.

Döll P, F Kaspar, I Alcamo 1999. Computation of Global Water Availability and Water Use at the Scale of Large Drainage Basins. Mathematische Geologie, 4:111–118.

Döll P, S Siebert 2002. Global Modeling of Irrigation Water Requirements. Water Resources Research, 38(4).

Dong Y. 2023. Irrigation Scheduling Methods: Overview and Recent Advances. Irrigation and Drainage - Recent Advances. IntechOpen. Available from: http://dx.doi.org/10.5772/intechopen.107386.

Dorado-Guerra DY, J Paredes-Arquiola, MA Pérez-Martín et al. 2023. Effect of Climate Change on the Water Quality of Mediterranean Rivers and Alternatives to Improve its Status. J. Environ. Manag, 348:119069.

Dotaniya ML, VD Meena, JK Saha et al. 2023. Reuse of Poor-Quality Water for Sustainable Crop Production in the Changing Scenario of Climate. Environ. Dev. Sustain, 25: 7345–7376.

Drake PL, RH Froend, PJ Franks 2013. Smaller, Faster Stomata: Scaling of Stomatal Size, Rate of Response, and Stomatal Conductance. J. Exp. Bot, 64:495–505.

Driscoll FG. 1986. Groundwater and wells (2nd ed., p. 1089). Johnson Division, St Paul.

Droogers P, WW Immerzeel, W Terink et al. 2012. Water Resources Trends in Middle East and North Africa towards 2050. Hydrol. Earth Syst. Sci, 16:3101–3114.

Du P, M Xu, R Li 2021. Impacts of Climate Change on Water Resources in the Major Countries along the Belt and Road. PeerJ, 9:e12201.

Dudley Ward N. 2024. The Groundwater Crisis: The Need for New Data to Inform Public Policy. Harvard Data Sci. Rev. 6(1). https://doi.org/10.1162/99608f92.67bb0dd2.

Du Plessis A 2022. Persistent Degradation: Global Water Quality Challenges and Required Actions. One Earth, 5(2):129–131.

Du Plessis JA, SG Kalima 2021. Modelling the Impact of Climate Change on the Flow of the Eerste River in South Africa. Phys. Chem. Earth (Pt A B,C), 124:103025.

Dupont BS, DL Allen. 2000. Revision of the Rainfall Intensity-Duration Curves for the Commonwealth of Kentucky. Ky Transp Cent, Coll Eng, University of Kentucky, Lexington.

Dutta P, AK Sarma 2021. Hydrological Modeling as a Tool For Water Resources Management of the Data-Scarce Brahmaputra Basin. J. Water Clim. Change, 12(1):152–165.

Eckstein G, F Sindico 2014. The Law of Transboundary Aquifers: Many Ways of Going Forward, but Only One Way of Standing Still. Review of European. Comp. Int. Environ. Law, 23:32–42.

EEA, 2018. European Waters—assessment of Status and Pressures. 2018. European Environment Agency Report No 7/2018. European Environment Agency, Copenhagen.

Ehtasham L, SH Sherani, F Nawaz 2024. Acceleration of the Hydrological Cycle and Its Impact on Water Availability Over Land: An Adverse Effect of Climate Change. Meteorol. Hydrol. Water Manag, 2(1):1–21.

Ehtasham L, SH Sherani, F Nawaz 2024. Acceleration of the Hydrological Cycle and Its Impact on Water Availability Over Land: An Adverse Effect of Climate Change. Meteorol. Hydrol. Water Manag, 12(1):1–21.

Eingrüber N, W Korres 2022. Climate Change Simulation and Trend Analysis of Extreme Precipitation and Floods in the Mesoscale Rur Catchment in Western Germany Until 2099 Using Statistical Downscaling Model (SDSM) and the Soil & Water Assessment Tool (SWAT model). Sci. Total Environ, 838:155775.

El-Ghzizel S, M Tahaikt, D Dhiba et al. 2021. Desalination in Morocco: Status and Prospects. Desalin. Water Treat, 231:1–15.

El-Nashar W, A Elyamany 2023. Adapting Irrigation Strategies to Mitigate Climate Change Impacts: A Value Engineering Approach. Water Resour. Manage, 37:2369–2386.

Eslamian S, S Parvizi, K Ostad-Ali-Askari et al. 2018. Water. In: Bobrowsky, P, B Marker (eds), Encyclopedia of Engineering Geology. Encyclopedia of Earth Sciences Series. Springer, Cham. https://doi.org/10.1007/978-3-319-12127-7_295-1.

Evans RG, EJ Sadler 2008. Methods and Technologies to Improve Efficiency of Water Use. Water Resour. Res, 44(7):1–15.

Fader M, S Shi, W von Bloh et al. 2016. Mediterranean Irrigation Under Climate Change: More Efficient Irrigation Needed to Compensate for Increases in Irrigation Water Requirements. Hydrol. Earth Syst. Sci, 20(2):953–973.

Falkenmark M 2003. Freshwater as Shared Between Society and Ecosystems: From Divided Approaches to Integrated Challenges. Philosophical Transactions of the Royal Society B, 358:2037–2049. https://doi.org/10.1098/rstb.2003.1386

Falkenmark M 2013. Growing Water Scarcity in Agriculture: Future Challenge to Global Water Security. Philosophical Transactions of the Royal Society A, 371:20120410. https://doi.org/10.1098/rsta.2012.0410

Falkenmark M, J Rockström 2006. The New Blue and Green Water Paradigm: Breaking New Ground for Water Resources Planning and Management. J. Water Resour. Plann. Manage, 132:129–132. https://doi.org/10.1061/(ASCE)0733-9496(2006)132: 3(129).

Fan X, X Cao, H Zhou et al. 2020. Carbon Dioxide Fertilization Effect on Plant Growth Under Soil Water Stress Associates with Changes in Stomatal Traits, Leaf Photosynthesis, and Foliar Nitrogen of Bell Pepper (*Capsicum annuum* L.). Environ Exp Bot, 179: 104203.

FAO. 2011. Climate Change, Water, and Food Security. Food Agric Organ Water Rep, 36:200.

FAO. 2012. Coping With Water Scarcity: An Action Framework for Agriculture and Food Security. FAO Water Reports 38, FAO.

FAO. 2019. Water Use in Livestock Production Systems and Supply Chains – Guidelines for Assessment (Version 1). Livestock Environmental Assessment and Performance (LEAP) Partnership.

FAO. 2023. The Impact of Disasters on Agriculture and Food Security 2023 – Avoiding and Reducing Losses through Investment in Resilience. FAO: Rome.

Farr TG, PA Rosen, E Caro et al. 2007. The Shuttle Radar Topography Mission. Rev. Geophys, 45:RG2004.

Fei S, R Wu, H Liu et al. 2025. Technological Innovations in Urban and Peri-Urban Agriculture: Pathways to Sustainable Food Systems in Metropolises. Horticulturae, 11(2):212.

Feng L, MA Raza, Z Li, et al. 2019. The Influence of Light Intensity and Leaf Movement on Photosynthesis Characteristics and Carbon Balance of Soybean. Front Plant Sci, 9, 1952. https://doi.org/10.3389/fpls.2018.01952.

Fernández JE 2017. Plant-Based Methods for Irrigation Scheduling of Woody Crops. Horticulturae, 3(2):35.

Fiaz A, G Rahman, HH Kwon 2025. Impacts of Climate Change on the South Asian Monsoon: A Comprehensive Review of Its Variability and Future Projection. J Hydroenviron Res, 59:100654.

Fienen MN, M Arshad 2016. The International Scale of the Groundwater Issue. In: Jakeman, AJ, O Barreteau, RJ Hunt, JD Rinaudo, A Ross (eds), Integrated Groundwater Management. Springer, Cham. https://doi.org/10.1007/978-3-319-23576-9_2

Fiorillo D, Z Kapelan, M Xenochristou et al. 2021. Assessing the Impact of Climate Change on Future Water Demand Using Weather Data. Water Resour Manage, 35:1449–1462. https://doi.org/10.1007/s11269-021-02789-4.

Fischer G, FN Tubiello, H van Velthuizen et al. 2007. Climate Change Impacts on Irrigation Water Requirements: Effects of Mitigation, 1990–2080. Technol Forecast Soc, 74: 1083–1107.

Fitton N, P Alexander, N Arnell et al. 2019. The Vulnerabilities of Agricultural Land and Food Production to Future Water Scarcity. Glob Environ Chang, 58:101944.

Flores-López F, SE Galaitsi, M Escobar et al. 2016. Modeling of Andean Páramo Ecosystems' Hydrological Response to Environmental Change. Water, 8(3):94.

Foley JA, N Ramankutty, KA Brauman et al. 2011. Solutions for a Cultivated Planet. Nature, 478(7369):337–42.

Fonjong L, RN Zama 2023. Climate Change, Water Availability, and The Burden of Rural Women's Triple Role in Muyuka, Cameroon. Global Environ Change, 82:102709.

Fowe T, H Karambiri, JE Paturel et al. 2015. Water Balance of Small Reservoirs in the Volta Basin: A Case Study of Boura Reservoir in Burkina FASO. Agric Water Manag, 152:99–109.

França ACF, RD Coelho, A da Silva Gundim et al. 2024. Effects of Different Irrigation Scheduling Methods on Physiology, Yield, and Irrigation Water Productivity of Soybean Varieties. Agric Water Manag, 293:108709.

Franke NA, H Boyacioglu, AY Hoekstra. 2013. Grey Water Footprint Accounting: Tier 1 Supporting Guidelines. Unesco-IHE, Delft.

Fry LM, DW Watkins, N Reents et al. 2012. Climate Change and Development Impacts on the Sustainability of Spring-Fed Water Supply Systems in the Alto Beni Region of Bolivia. J Hydrol, 468–469:120–129.

Fukase E, W Martin 2020. Economic Growth, Convergence, and World Food Demand and Supply. World Dev, 132:104954.

Gabr ME 2023. Impact of Climatic Changes on Future Irrigation Water Requirement in the Middle East and North Africa's Region: A Case Study of Upper Egypt. Appl Water Sci, 13:158.

Gain AK, C Giupponi, Y Wada 2016. Measuring Global Water Security towards Sustainable Development Goals. Res Lett, 11:124015. https://doi.org/10.1088/1748-9326/11/12/124015.

Galdies C 2022. A Multidecadal Analysis of Malta's Climate Trends and Extreme Events, 1952–2022. ResearchGate, 2022:1–11.

Gan Y, KHM Siddique, NC Turner et al. 2013. Ridge-Furrow Mulching Systems-An Innovative Technique for Boosting Crop Productivity in Semiarid Rain-Fed Environments. Adv Agron, 118:429–476.

Ganti V, AD Sarma. 2013. Data Cleaning: A Practical Perspective. Synthesis Lectures on Data Management, 85. Morgan & Claypool Publishers: California.

Gao T, X Wang, D Wei et al. 2021. Transboundary Water Scarcity under Climate Change. J Hydrol, 598:126453.

García JEA, B Yazici, A Richa et al. 2024. Digitalising Governance Processes and Water Resources Management to Foster Sustainability Strategies in the Mediterranean Agriculture. Environ Sci Policy, 158:103805.

Garcia-Ruiz JM, JI Lopez-Moreno, SM Vicente-Serrano et al. 2011. Mediterranean Water Resources in a Global Change Scenario. Earth Sci Rev, 105:121–139.

Garg KK, V Akuraju, KH Anantha et al. 2022. Identifying Potential Zones for Rainwater Harvesting Interventions for Sustainable Intensification in the Semi-Arid Tropics. Sci Rep, 12(1):3882.

Gebrehiwot K, M Gebrewahid 2016. The Need for Agricultural Water Management in Sub-Saharan Africa. J Water Resour Protect, 8:835–843. https://doi.org/10.4236/jwarp.2016.89068

Gebremedhin T, GG Haile, TG Gebremicael et al. 2023. Balancing Crop Water Requirements Through Supplemental Irrigation Under Rainfed Agriculture in a Semi-Arid Environment. Heliyon, 9(8):e18727.

Gee GW, D Hillel 1988. Groundwater Recharge in Arid Regions: Review and Critique of Estimation Methods. Hydrol Process, 2:255–266.

Geerts S, D Raes 2009. Deficit Irrigation as an On-Farm Strategy to Maximize Crop Water Productivity in Dry Areas. Agric Water Manag, 96:1275–1284.

Gerveni M, AFT Avelino, S Dall'erba 2020. Drivers of Water Use in the Agricultural Sector of the European Union 27. Environ. Sci. Technol, 54(15):9191–9199.

Gettelman A, RB Rood. 2016. Simulating Terrestrial Systems. In: Demystifying Climate Models. Earth Systems Data and Models, vol. 2, Springer, Berlin, Heidelberg. https://doi.org/10.1007/978-3-662-48959-8_7.

Giménez C, M Gallardo, RB Thompson. 2013. Plant–Water Relations. Reference Module in Earth Systems and Environmental Sciences. Elsevier. https://doi.org/10.1016/B978-0-12-409548-9.05257-X

Giri A, VK Bharti, S Kalia et al. 2020. A Review on Water Quality and Dairy Cattle Health: A Special Emphasis on High-Altitude Region. Appl Water Sci, 10:79.

Githui F, W Gitau, F Mutua et al. 2009. Climate Change Impact on SWAT Simulated Streamflow in Western Kenya. Int J Climatol, 29:1823–1834.

Global Commission on Adaptation. 2019. Adapt Now: A Global Call for Leadership on Climate Resilience. https://gca.org/wp-content/uploads/2019/09/GlobalCommission_Report_FINAL.pdf.

Gómez-Ramos A, I Blanco-Gutiérrez, M Ballesteros-Olza et al. 2024. Are Non-Conventional Water Resources the Solution for the Structural Water Deficit in Mediterranean Agriculture? The Case of the Segura River Basin in Spain. Water, 16(7):929.

Gong Z, F Gao, X Chang et al. 2025. A Review of Interactions between Irrigation and Evapotranspiration. Ecol Indic, 169:112870.

Gornall J, R Betts, E Burke et al. 2010. Implications of Climate Change for Agricultural Productivity in The Early Twenty-First Century. Philos Trans R Soc B: Biol Sci, 365(1554):2973–2989.

Goswami BN, V Krishnamurthy, H Annamalai 1999. A Broad-Scale Circulation Index for the Interannual Variability of the Indian Summer Monsoon. Q J R Meteorol Soc, 125:611–633.

Gramlich A, S Stoll, C Stamm et al. 2018. Effects of Artificial Land Drainage on Hydrology, Nutrient and Pesticide Fluxes from Agricultural Fields–A Review. Agric Ecosyst Environ, 266:84–99.

Groombridge B, M Jenkins. 1998. Freshwater Biodiversity: A Preliminary Global Assessment. WCMC–World Conservation Press.

Grossi G, P Goglio, A Vitali et al. 2018. Livestock and Climate Change: Impact of Livestock on Climate and Mitigation Strategies. Anim Front, 9(1):69–76. https://doi.org/10.1093/af/vfy014.

Gruère G, M Shigemitsu, S Crawford. 2020. Agriculture and Water Policy Changes: Stocktaking and Alignment with OECD and G20 Recommendations. OECD Food, Agriculture and Fisheries Papers, No. 144, OECD Publishing, Paris. https://doi.org/10.1787/f35e64af-en.

Gründemann GJ, N van de Giesen, L Brunner et al. 2022. Rarest Rainfall Events Will See the Greatest Relative Increase in Magnitude under Future Climate Change. Commun Earth Environ, 3:235.

Grusson Y, F Anctil, S Sauvage et al. 2018. Coevolution of Hydrological Cycle Components under Climate Change: The Case of the Garonne River in France. Water, 10(12):1870.

Grusson Y, I Wesström, E Svedberg et al. 2021. Influence of Climate Change on Water Partitioning in Agricultural Watersheds: Examples from Sweden. Agric Water Manag, 249:106766.

Gu D, K Andreev, ME Dupre 2021. Major Trends in Population Growth Around the World. China CDC Wkly, 3(28):604–613.

Gullino ML, M Pugliese, G Gilardi et al. 2018. Effect of Increased CO_2 and Temperature on Plant Diseases: A Critical Appraisal of Results Obtained in Studies Carried Out Under Controlled Environment Facilities. J Plant Pathol, 100:371–389. https://doi.org/10.1007/s42161-018-0125-8.

Guo H, S Li 2024. A Review of Drip Irrigation's Effect on Water, Carbon Fluxes, and Crop Growth in Farmland. Water, 16(15):2206.

Gupta N, E Humphreys, P Eberbach et al. 2021. Effects of Tillage and Mulch on Soil Evaporation in a Dry Seeded Rice-Wheat Cropping System. Soil Tillage Res, 209: 104976.

Gurovich LA, L Fernando Riveros 2019. Agronomic Operation and Maintenance of Field Irrigation Systems', Irrigation - Water Productivity and Operation, Sustainability and Climate Change. IntechOpen: London. https://doi.org/10.5772/intechopen.84997

Hadour A, G Mahé, M Meddi 2020. Watershed Based Hydrological Evolution Under Climate Change Effect: An Example from North-Western Algeria. J Hydrol Reg Stud, 28:100671.

Haidu I, MM Nistor 2020. Long-Term Effect of Climate Change on Groundwater Recharge in the Grand Est Region of France. Meteorol Appl, 27(1):e1796

Haile GG, Q Tang, KW Reda et al. 2024. Projected Impacts of Climate Change on Global Irrigation Water Withdrawals. Agric Water Manag, 305:109144.

Haj-Amor Z, T Araya, DG Kim et al. 2022. Soil Salinity and its Associated Effects on Soil Microorganisms, Greenhouse Gas Emissions, Crop Yield, Biodiversity and Desertification: A Review. Sci Total Environ, 843:156946. https://doi.org/10.1016/j.scitotenv.2022.156946.

Haj-Amor Z, MK Ibrahimi, N Feki et al. 2016. Soil Salinisation and Irrigation Management of Date Palms in a Saharan Environment. Environ Monit Assess, 188:497.

Haj-Amor Z, D-G Kim, S Bouri. 2023. Sustainable Agriculture: Adaptation Strategies to Address Climate Change by 2050 (1st ed.). CRC Press: Florida. https://doi.org/10.1201/9781003404194

Hamad K, N Khan, Z Khan 2025. Water and Heat Resource Utilization Influence Cotton Yield Through Sowing Date Optimization Under Varied Climate. Agric Water Manag, 313:109491.

Hamed Y, R Hadji, B Redhaounia et al. 2018. Climate Impact on Surface and Groundwater in North Africa: A Global Synthesis of Findings and Recommendations. Euro-Mediterr J Environ Integr. 3:25.

Hamidov A, K Helming, D Balla 2016. Impact of Agricultural Land Use in Central Asia: A Review. Agron Sustain Dev, 36:6.

Hammouri N, J Adamowski, M Freiwan et al. 2017. Climate Change Impacts on Surface Water Resources in Arid and Semi-Arid Regions: A Case Study in Northern Jordan. Acta Geod Geophys, 52:141–156.

Hanjra MA, ME Qureshi 2010. Global Water Crisis and Future Food Security in an Era of Climate Change. Food Policy, 35(5):365–377. https://doi.org/10.1016/j.foodpol.2010.05.002.

Harbaugh AW. 2005. MODFLOW-2005, U.S. Geological Survey Modular Ground-Water Model—The Groundwater Flow Process. U.S. Geological Survey Techniques and Methods 6-A16.

Hariadi MH, G van der Schrier, GJ Steeneveld et al. 2024. A High-Resolution Perspective of Extreme Rainfall and River Flow Under Extreme Climate Change in Southeast Asia. Hydrol Earth Syst Sci, 28:1935–1956.

Harrington C. (2013). Fluid Identities: Toward a Critical Security of Water. Electronic Thesis and Dissertation Repository. Paper 1716.

Hasan F, P Medley, J Drake et al. 2024. Advancing Hydrology Through Machine Learning: Insights, Challenges, and Future Directions Using The CAMELS, Caravan, GRDC, CHIRPS, PERSIANN, NLDAS, GLDAS, and GRACE Datasets. Water, 16(13):1904.

Hashmi HA, AO Belgacem, M Behnassi et al. 2021. Impacts of Climate Change on Livestock and Related Food Security Implications—Overview of the Situation in Pakistan and Policy Recommendations. In: Behnassi, M, M Barjees Baig, M El Haiba, MR Reed (eds), Emerging Challenges to Food Production and Security in Asia, Middle East, and Africa. Springer, Cham. https://doi.org/10.1007/978-3-030-72987-5_8.

Hasselquist EM, W Lidberg, RA Sponseller et al. 2018. Identifying and Assessing the Potential Hydrological Function of Past Artificial Forest Drainage. Ambio, 47:546–556.

He L, L Rosa 2023. Solutions to Agricultural Green Water Scarcity Under Climate Change. PNAS Nexus, 2(4):117.

Heinke J, M Lannerstad, D Gerten et al. 2020. Water use in Global Livestock Production – Opportunities and Constraints for Increasing Water Productivity. Water Resour.Res, 56:e2019WR026995.

Helmecke M, E Fries, C Schulte 2020. Regulating Water Reuse for Agricultural Irrigation: Risks Related to Organic Micro-Contaminants. Environ Sci Eur, 32:4.

Hengsdijk H, AAMFR Smit, JG Conijn, et al. 2014. Agricultural Crop Potentials and Water Use in East Africa. (Report/Plant Research International; No. 555). Plant Research International. https://edepot.wur.nl/304249.

Hermes Ulises RS, F Montiel Aida Lucia, OB Alma Delia et al. 2022. Impacts of Climate Change on the Water Sector in Mexico. Asian J Environ Ecol, 17(2):37–57.

Herrera PA, MA Marazuela, T Hofmann 2022. Parameter Estimation and Uncertainty Analysis in Hydrological Modeling. Wiley Interdiscip Rev Water, 9(1):e1569.

Hikouei IS, KN Eshleman, BH Saharjo et al. 2023. Using Machine Learning Algorithms to Predict Groundwater Levels in Indonesian Tropical Peatlands. Sci Total Environ, 857:159701.

Hirpa FA, E Dyer, R Hope 2018. Finding Sustainable Water Futures in Data-Sparse Regions Under Climate Change: Insights from the Turkwel River Basin, Kenya. J Hydrol Reg Stud, 19:124–135.

Hiscock K, R Sparkes, A Hodgson 2011. Evaluation of Future Climate Change Impacts on European Groundwater Resources. In: Climate Change Effects on Groundwater Resources: A Global Synthesis of Findings and Recommendations (pp. 351–365). CRC Press: Florida.

HLPE. 2014. Sustainable Fisheries and Aquaculture for Food Security and Nutrition. A Report by the High-Level Panel of Experts on Food Security and Nutrition of the Committee on World Food Security. FAO: Rome.

Hoekstra AY, AK Chapagain, MM Aldaya, et al. 2011. The Water Footprint Assessment Manual. Earthscan: London and Washington, DC.

Hoekstra AY, MM Mekonnen. 2011. Global Water Scarcity: Monthly Blue Water Footprint Compared to Blue Water Availability for the World's Major River Basins. Value of Water Research Report Series No. 53, UNESCO-IHE, Delft, The Netherlands.

Hoekstra AY, MM Mekonnen 2012. The Water Footprint of Humanity. Proc Natl Acad Sci USA, 109:3232–3237. https://doi.org/10.1073/pnas.1109936109.

Hoekstra AY, MM Mekonnen, AK Chapagain et al. 2012. Global Monthly Water Scarcity: Blue Water Footprints Versus Blue Water Availability. PLoS One, 7(2):e32688. https://doi.org/10.1371/journal.pone.0032688.

Homobono T, MH Guimarães, C Esgalhado et al. 2022. Water Governance in Mediterranean Farming Systems through the Social-Ecological Systems Framework—An Empirical Case in Southern Portugal. Land, 11(2):178.

Hoque MAA, B Pradhan, N Ahmed et al. 2021. Agricultural Drought Risk Assessment of Northern New South Wales, Australia Using Geospatial Techniques. Sci Total Environ, 756:143600.

Horton P, B Schaefli, M Kauzlaric 2022. Why do We have so Many Different Hydrological Models? A Review Based on the Case of Switzerland. Wiley Interdiscip Rev Water, 9(1):e1574.

Hossen MA, J Connor, F Ahammed 2023. How to Resolve Transboundary River Water Sharing Disputes. Water, 15(14):2630.

Howden M, S Schroeter, S Crimp et al. 2014. The Changing Roles of Science in Managing Australian Droughts: An Agricultural Perspective. Wea Climate Extremes, 3:80–89.

Hrdinka T, O Nový, E Hanslík et al. 2012. Possible Impacts of Floods and Droughts on Water Quality. J Hydro Environ Res, 6:145–150.

Hristov J, J Barreiro-Hurle, G Salputra et al. 2021. Reuse of Treated Water in European Agriculture: Potential to Address Water Scarcity Under Climate Change. Agric Water Manag, 251:106872.

Huang F, T Du, S Wang et al. 2019. Current Situation and Future Security of Agricultural Water Resources in North China. Strat Study of CAE, 21(5):28–37. https://doi.org/10.15302/J-SSCAE-2019.05.024

Huang S, S Eisner, WK Wong et al. 2025. The Potential Impacts of Climate and Forest Changes on Streamflow for Micro-, Meso- and Macro-Scale Catchments in Norway. J Hydrol Reg Stud, 57:102147.

Huang Z, M Hejazi, X Li, Q Tang, C Vernon, G Leng, Y Liu, P Döll, S Eisner, D Gerten, N Hanasaki, Y Wada 2018. Reconstruction of Global Gridded Monthly Sectoral Water Withdrawals for 1971–2010 and Analysis of Their Spatiotemporal Patterns. Hydrol Earth Syst Sci, 22:2117–2133.

Huang Z, M Hejazi, Q Tang et al. 2019a. Global Agricultural Green and Blue Water Consumption Under Future Climate and Land Use Changes. J Hydrol, 574:242–256.

Hung WC, C Hwang, SH Lin et al. 2024. Exploring Groundwater Depletion and Land Subsidence Dynamics in Taiwan's Choushui River Alluvial Fan: Insights from Integrated GNSS and Hydrogeological Data Analysis. Front Earth Sci, 12:1370626.

Hurlbert M, P Mussetta 2016. Creating Resilient Water Governance for Irrigated Producers in Mendoza, Argentina. Environ. Sci. Policy, 58:83–94.

Huss M, R Hock 2018. Global-Scale Hydrological Response to Future Glacier Mass Loss. Nat Clim Change, 8:135–140.

Hussain MS, HF Abd-Elhamid, AA Javadi et al. 2019. Management of Seawater Intrusion in Coastal Aquifers: A Review. Water, 11(12):2467.

Ibarrola-Rivas MJ, R Granados-Ramírez, S Nonhebel 2017. Is the Available Cropland and Water Enough for Food Demand? A Global Perspective of the Land-Water-Food Nexus. Adv Water Resour, 110:476–483.

IDMP. 2022. Drought and Water Scarcity. WMO No. 1284. Global Water Partnership, Stockholm, Sweden and World Meteorological Organization, Geneva, Switzerland.

Idrizovic D, V Pocuca, MV Mandic et al. 2020. Impact of Climate Change on Water Resource Availability in a Mountainous Catchment: A Case Study of The Toplica River Catchment, Serbia. J Hydrol, 587:124992.

Ierna A 2023. Water Management in Potato. In: Calis Kan, ME, A Bakhsh, K Jabran (eds), Potato Production Worldwide (pp. 87–100). Academic Press: New York.

Igbadun HE, BA Salim, A Tarimo et al. 2008. Effects of Deficit Irrigations Scheduling on Yields and Soil Water Balance of Irrigated Maize. Irrig Sci, 27:11–23.

Immerzeel WW, AF Lutz, M Andrade et al. 2020. Importance and Vulnerability of the World's Water Towers. Nature, 577(7790):364–369.

Ingrao C, R Strippoli, G Lagioia et al. 2023. Water scarcity in Agriculture: An Overview of Causes, Impacts, and Approaches for Reducing the Risks. Heliyon, 9(8):e18507.

IPCC. 2021. Climate Change 2021: The Physical Science Basis. Contribution of Working Group I to The Sixth Assessment Report of the Intergovernmental Panel on Climate Change, Vol 2.

Irmak S, LO Odhiambo, WL Kranz, et al. 2011. Irrigation Efficiency and Uniformity and Crop Water Use Efficiency. Publication EC732. University of Nebraska-Lincoln Extension: Nebraska.

Isukuru EJ, JO Opha, OW Isaiah et al. 2024. Nigeria's Water Crisis: Abundant Water, Polluted Reality. Cleaner Water, 2:100026.

Jacobsen D, AM Milner, LE Brown et al. 2012. Biodiversity Under Threat in Glacier-Fed River Systems. Nat Clim Change, 2:361–364.

Jacoby HG, G Mansuri, F Fatima 2021. Decentralizing Corruption: Irrigation Reform in Pakistan. J Public Econ, 202:104499.

Jain R, P Kishore, DK Singh 2019. Irrigation in India: Status, Challenges and Options. J Soil Water Conserv, 18(4):354–363.

Jaiswal RK, S Ali, B Bharti 2020. Comparative Evaluation of Conceptual And Physical Rainfall–Runoff Models. Appl Water Sci, 10:48.

Janjua S, I Hassan, S Muhammad et al. 2021. Water Management in Pakistan's Indus Basin: Challenges and Opportunities. Water Policy, 23(6):1329–1343.

Jasechko S, H Seybold, D Perrone et al. 2024. Rapid Groundwater Decline and Some Cases of Recovery in Aquifers Globally. Nature, 625:715–721.

Javeline D, R Orttung, G Robertson et al. 2024. Russia in a Changing Climate. WIREs Climate Change, https://doi.org/10.1002/wcc.872.

Jeyrani F, S Morid, R Srinivasan 2021. Assessing Basin Blue–Green Available Water Components Under Different Management and Climate Scenarios Using SWAT. Agric Water Manage, 256:107074.

Jian S, X Cheng, T Wang et al. 2025. The Driving Factors of Water Use and Its Decoupling Relationship with Economic Development—A Multi-Sectoral Perspective in the Nine Provinces of the Yellow River Basin. J Hydrol Reg Stud, 59:102338.

Jiang B, CP Wong, F Lu et al. 2014. Drivers of Drying on the Yongding River in Beijing. J Hydrol, 519:69–79.

Jiang L, L Yu 2019. Analyzing Land Use Intensity Changes Within and Outside Protected Areas Using ESA CCI-LC Datasets. Glob Ecol Conserv, 20:e00789.

Johansson EL, M Fader, JW Seaquist et al. 2016. Green and Blue Water Demand from Large-Scale Land Acquisitions in Africa. PNAS, 113:11471–11476.

Jones E, MTH van Vliet 2018. Drought Impacts on River Salinity in the Southern Us: Implications for Water Scarcity. Sci Total Environ or STOTEN, 644:844–853.

Jonnalagadda SB, G Mhere 2001. Water Quality of the Odzi River in the Eastern Highlands of Zimbabwe. Water Res, 35:2371–2376.

Joshi SK, S Gupta, R Sinha et al. 2021. Strongly Heterogeneous Patterns of Groundwater Depletion in Northwestern India. J Hydrol, 598:126492.

Jury WA, H Vaux 2005. The Role of Science in Solving the World's Emerging Water Problems. Proc Natl Acad Sci USA, 102:15715–15720.

Kakabayev A, B Yessenzholov, A Khussainov et al. 2023. The Impact of Climate Change on the Water Systems of the Yesil River Basin in Northern Kazakhstan. Sustainability, 15(22):15745.

Kannan N, SM White, F Worrall et al. 2007. Sensitivity Analysis and Identification of the Best Evapotranspiration and Runoff Options for Hydrological Modelling in SWAT-2000. J Hydrol, 332(3–4):456–466.

Katerji N, M Mastrorilli, G Rana 2008. Water use Efficiency of Crops Cultivated in The Mediterranean Region: Review and Analysis. Eur J Agron, 28:493–507. https://doi.org/10.1016/j.eja.2008.03.001.

Ke J, H Blum 2021. A Panel Analysis of Groundwater Use in California. J Clean Prod, 326:12.

Kelly S, R Cunningham, R Plant, K Maras, 2019, Water Scarcity Risk for Australian Farms and the Implications for the Financial Sector. Institute for Sustainable Futures, UTS.

Khaliq P, S Razavi, EGR Davies et al. 2023. Assessment of Blue Water-Green Water Interchange Under Extreme Warm and Dry Events Across Different Ecohydrological Regions of Western Canada. J Hydrol, 625:130105.

Khapte PS, P Kumar, U Burman et al. 2019. Deficit Irrigation in Tomato: Agronomical and Physio-Biochemical Implications. Sci Hortic, 248:256–264.

Kharazi P, MR Yazdani, P Khazealpour 2019. Suitable Identification of Underground Dam Locations, Using Decision-Making Methods in a Semi-Arid Region of Iranian Semnan Plain. Groundw Sustain Dev, 9:100240.

Kharraz E, A El-Sadek, N Ghaffour et al. 2012. Water Scarcity and Drought in WANA Countries. Procedia Eng, 33:14–29.

Khlifi S, M Ameur, N Mtimet et al. 2010. Impacts of Small Hill Dams on Agricultural Development of Hilly Land in the Jendouba Region of Northwestern Tunisia. Agric Water Manag, 97:50–56.

Kim B, BF Sanders, JS Famiglietti et al. 2015. Urban Flood Modeling with Porous Shallow-Water Equations: A Case Study of Model Errors in the Presence of Anisotropic Porosity. J Hydrol, 523:680–692.

Kim H, S Kim, J Jeon et al. 2020. Effects of Irrigation with Desalinated Water on Lettuce Grown under Greenhouse in South Korea. Appl Sci, 10(7):2207.

Kim W, T Iizumi, M Nishimori 2019. Global Patterns of Crop Production Losses Associated with Droughts from 1983 to 2009. J.Appl Meteor Climatol, 58:1233–1244.

Koech R, P Langat 2018. Improving Irrigation Water Use Efficiency: A Review of Advances, Challenges and Opportunities in the Australian Context. Water, 10(12):1771.

Koehler T, FJP Wankmüller, W Sadok et al. 2023. Transpiration Response to Soil Drying Versus Increasing Vapor Pressure Deficit in Crops: Physical and Physiological Mechanisms and Key Plant Traits. J Exp Bot, 74(16):4789–4807.

Konapala G, AK Mishra, Y Wada et al. 2020. Climate Change Will Affect Global Water Availability Through Compounding Changes in Seasonal Precipitation and Evaporation. Nat Commun, 11:3044.

Konapala G, AK Mishra, Y Wada 2020. Climate Change will Affect Global Water Availability Through Compounding Changes in Seasonal Precipitation and Evaporation. Nat Commun, 11(1):1–10.

Konzmann M, D Gerten, J Heinke 2013. Climate Impacts on Global Irrigation Requirements Under 19 Gcms, Simulated with a Vegetation and Hydrology Model. Hydrol Sci J, 58(1):1–18.

Körner C. 2013. Plant–Environment Interactions. In: Strasburger's Plant Sciences. Springer, Berlin, Heidelberg. https://doi.org/10.1007/978-3-642-15518-5_12.

Korzun, VI, ed. 1974. World Water Balance and Water Resources of the Earth. Leningrad, Hydrometeoizdat.

Kourakos G, HE Dahlke, T Harter 2019. Increasing Groundwater Availability and Seasonal Base Flow Through Agricultural Managed Aquifer Recharge in An Irrigated Basin. Water Resour Res, 55:7464–7492.

Kourgialas NN 2021. Hydroclimatic Impact on Mediterranean Tree Crops Area–Mapping Hydrological Extremes (Drought/Flood) Prone Parcels. J Hydrol, 596:125684.

Kovats RS, R Valentini, LM Bouwer et al. 2014. Europe. In: Climate Change 2014: Impacts, Adaptation, and Vulnerability. Part B: Regional Aspects. Contribution of Working Group II to the Fifth Assessment Report of the Intergovernmental Panel on Climate Change (pp. 1267–1326). Cambridge University Press: Cambridge.

Kuchimanchi BR, R Ripoll-Bosch, FA Steenstra et al. 2023. The Impact of Intensive Farming Systems on Groundwater Availability in Dryland Environments: A Watershed Level Study from Telangana, India. Curr Res Environ Sustain, 5:100198.

Kumar A, S Kanga, AK Taloor et al. 2021. Surface Runoff Estimation of Sind River Basin Using Integrated SCS-CN and GIS Techniques. HydroResearch, 4:61–74.

Kumar TJR, A Balasubramanian, RS Kumar et al. 2016. Assessment of Groundwater Potential Based on Aquifer Properties of Hard Rock Terrain in the Chittar–Uppodai Watershed, Tamil Nadu, India Appl Water Sci, 6:179–186.

Kummu M, J Guillaume, H de Moel et al. 2016. The World's Road to Water Scarcity: Shortage and Stress in the 20th Century and Pathways Towards Sustainability. Sci Rep, 6:38495.

Kusangaya S, ML Warburton, EA van Garderen et al. 2014. Impacts of Climate Change on Water Resources in Southern Africa: A Review. Phys Chem Earth, 67–69:47–54.

Kuzma S, L Saccoccia, M Chertock 2023. 25 Countries, Housing One-Quarter of the Population, Face Extremely High-Water Stress. World Resources Institute (WRI), Washington, DC, USA.

Laamari A, H Laamrani, S Khoali 2022. Smart Water Governance in Moroccan Agriculture: New Science and Policy Collaboration and Partnership. Open J Soc Sci, 10:11–27.

La Jeunesse I, C Cirelli, D Aubin et al. 2016. Is Climate Change a Threat for Water uses in the Mediterranean Region? Results from a Survey at Local Scale. Sci Total Environ, 543(Part B):981–9.

Lakshmi V, J Fayne, J Bolten 2018. A Comparative Study of Available Water in the Major River Basins of The World. J Hydrol, 567:510–532.

Lall U, L Josset, T Russo 2020. A Snapshot of the World's Groundwater Challenges. Annu Rev Environ Resour, 45:171–94.

Lee E, K Jung, R Jayakumar 2024. Status of Water Security in East and Southeast Asia. In: Lee, E., Böer, B., Surendra, L., Chun, J.A., Taniguchi, M. (eds) The Water, Energy, and Food Security Nexus in Asia and the Pacific. Water Security in a New World. Springer, Cham. https://doi.org/10.1007/978-3-031-12495-2_2.

Lee J, D Perera, T Glickman et al. 2020. Water-Related Disasters and Their Health Impacts: A Global Review. Prog Disaster Sci, 8:100123.

Leong Tan M, WP Gassman, J James Haywood 2020. A Review of SWAT Applications, Performance and Future needs For Simulation of Hydro-Climatic Extremes. Adv Water Res, 143:103662.

Leta OT, W Bauwens 2018. Assessment of the Impact of Climate Change on Daily Extreme Peak and Low Flows of Zenne Basin in Belgium. Hydrol, 5(3):38.

Levidow L, D Zaccaria, R Maia et al. 2014. Improving Water-Efficient Irrigation: Prospects and Difficulties Of Innovative Practices. Agric. Water Manag, 146:84–94.

Levintal EML, Y Kniffin, N Ganot et al. 2023. Agricultural Managed Aquifer Recharge (Ag-Mar) – A Method for Sustainable Groundwater Management: A Review. Crit Rev Environ Sci Technol, 53(3):291–314.

Li B, W Xiao, Y Wang et al. 2018. Impact of Land Use/Cover Change on The Relationship between Precipitation and Runoff in Typical Area. J Water Clim Change, 9(2):261–274.

Li X, W Shi, K Broughton et al. 2020. Impacts of Growth Temperature, Water Deficit And Heatwaves on Carbon Assimilation and Growth Of Cotton Plants (*Gossypium hirsutum* L.). Environ Exp Bot, 179:104204.

Li Y, W Mi, L Ji et al. 2023. Urbanization and Agriculture Intensification Jointly Enlarge the Spatial Inequality of River Water Quality. Sci Total Environ, 878:162559.

Li YJ, YJ Ding, DH Shangguan et al. 2019. Regional Differences in Global Glacier Retreat from 1980 to 2015. Adv Clim Change Res, 10(4):203–213.

Liang X, DP Lettenmaier, EF Wood et al. 1994. A Simple Hydrologically Based Model of Land Surface Water and Energy Fluxes for General Circulation Models. J Geophys Res, 99(D7):14415–14428.

Liangliang GAO, LI Daoliang 2014. A Review of Hydrological/Water-Quality Models. Front Agr Sci Eng, 1(4):267–276.

Lin BB 2007. Agroforestry Management as an Adaptive Strategy Against Potential Microclimate Extremes in Coffee Agriculture. Agric For Meteorol, 150(4):510–518.

Lin F, X Li, N Jia et al. 2023. The Impact of Russia-Ukraine Conflict on Global Food Security. Glob Food Sec, 36:100661.

Linderholm HW 2006. Growing Season Changes in the Last Century. Agric For Meteorol, 137:1–14.

Liu J, Z Fu, W Liu 2023. Impacts of Precipitation Variations on Agricultural Water Scarcity Under Historical and Future Climate Change. J Hydrol, 617:128999.

Liu J, T Hu, L Fang et al. 2019. CO_2 Elevation Modulates the Response of Leaf Gas Exchange to Progressive Soil Drying in Tomato Plants. Agric Meteorol, 268:181–188.

Liu J, Q Liu, H Yang 2016. Assessing Water Scarcity by Simultaneously Considering Environmental Flow Requirements, Water Quantity, and Water Quality. Ecol Indic, 60:434–441.

Liu L, L Zhang, Q Zhang et al. 2024. A Warming-Induced Glacier Reduction Causes Lower Streamflow in the Upper Tarim River Basin. J Hydrol Reg Stud, 53:101802.

Liu R, L Xing, G Zhou et al. 2017a. What is Meat in China? Anim. Frontiers, 7(4):53–56.

Liu W, X Liu, H Yang et al. 2022a. Global Water Scarcity Assessment Incorporating Green Water in Crop Production. Water Resour Res, 58:e2020WR028570.

Liu X, W Liu, Q Tang et al. 2022b. Global Agricultural Water Scarcity Assessment Incorporating Blue and Green Water Availability Under Future Climate Change. Earth's Future, 10:e2021EF002567.

Liu Z, Y Wang, Z Xu et al. 2017b. Conceptual Hydrological Models. In: Duan, Q., Pappenberger, F., Thielen, J., Wood, A., Cloke, H., Schaake, J. (eds) Handbook of Hydrometeorological Ensemble Forecasting. Springer, Berlin, Heidelberg.

Loi NK, VNQ Tram, NTT Au 2020. Climate Change Impacts on Hydrology in the Dak B'la Watershed, Central Highland Vietnam Based on SWAT model. Eur J Clim Ch, 02(01):22–31.

Loladze I 2014. Hidden Shift of The Ionome of Plants Exposed to Elevated CO_2 Depletes Minerals at the Base of Human Nutrition. eLife, 3:e02245.

López-Ballesteros A, J Senent-Aparicio, C Martínez et al. 2020. Assessment of Future Hydrologic Alteration Due to Climate Change in the Aracthos River Basin (NW Greece). Sci Total Environ, 733:139299.

Lovrinović I, V Srzić, I Aljinović 2023. Characterization of Seawater Intrusion Dynamics Under the Influence of Hydro-Meteorological Conditions, Tidal Oscillations and Melioration System Operative Regimes to Groundwater in Neretva Valley Coastal Aquifer System. J Hydrol Reg Stud, 46:101363.

Luo D, J He 2023. Just Commons: Governance of Irrigation Water in World Heritage Rice Terraces, Southwest China. Int. J. Commons, 17(1):141–154. Available at: https://doi.org/10.5334/ijc.1203.

Luo H, X Nong, H Xia et al. 2024. Integrating Water Quality Index (WQI) and Multivariate Statistics for Regional Surface Water Quality Evaluation: Key Parameter Identification and Human Health Risk Assessment. Water, 16(23):3412.

Luo H, Z Wang, C He et al. 2024. Future Changes in South Asian Summer Monsoon Circulation Under Global Warming: Role of the Tibetan Plateau Latent Heating. npj Clim Atmos Sci, 7:103.

Luo J, E Straffelini, M Bozzolan et al. 2024. Saltwater Intrusion in the Po River Delta (Italy) During Drought Conditions: Analyzing Its Spatio-Temporal Evolution and Potential Impact on Agriculture. Int Soil Water Conserv Res, 12:714–725.

Lupia F, V Baiocchi, K Lelo et al. 2017. Exploring Rooftop Rainwater Harvesting Potential for Food Production in Urban Areas. Agriculture, 7(6):46.

Lutz AF, WW Immerzeel, AB Shrestha et al. 2014. Consistent Increase in High Asia's Runoff Due to Increasing Glacier Melt and Precipitation. Nat Clim Change, 4:587–592.

Lutz AF, WW Immerzeel, C Siderius et al. 2022. South Asian Agriculture Increasingly Dependent on Meltwater and Groundwater. Nat Clim Chang, 12:566–573.

Lyra A, A Loukas, P Sidiropoulos et al. 2024. Climate Change Impacts on Nitrate Leaching and Groundwater Nitrate Dynamics Using a Holistic Approach and Med-CORDEX Climatic Models. Water, 16(3):465.

Maansi R, M Jindal, Wats 2022. Evaluation of Surface Water Quality Using Water Quality Indices (WQIs) in Lake Sukhna, Chandigarh, India. Appl Water Sci, 12(2): 1 -14.

Mabhaudhi T, L Nhamo, S Mpandeli et al. 2021. Enhancing Crop Water Productivity Under Increasing Water Scarcity in South Africa. In: Ting, DSK, JA Stagner (eds), Climate Change Science (pp. 1–18). Elsevier: Amsterdam, Netherlands.

Mahajan PV, FAR Oliveira, I Macedo 2008. Effect of Temperature and Humidity on the Transpiration Rate of the Whole Mushrooms. J Food Eng, 84(2):281–288. https://doi.org/10.1016/j.jfoodeng.2007.05.021

Mahmoud SH, TY Gan, RP Allan et al. 2022. Worsening Drought of Nile Basin Under Shift in Atmospheric Circulation, Stronger Enso and Indian Ocean Dipole. Sci Rep, 12(1):8049.

Malek K, JC Adam, CO Stockle et al. 2018. Climate Change Reduces Water Availability for Agriculture by Decreasing Non-Evaporative Irrigation Losses. J Hydrol, 561:444–460.

Malekian A, D Hayati, N Aarts 2017. Conceptualizations of Water Security in the Agricultural Sector: Perceptions, Practices, and Paradigms. J Hydrol, 544:224–232.

Mall RK, RK Srivastava, T Banerjee et al. 2019. Disaster Risk Reduction Including Climate Change Adaptation Over South Asia: Challenges and Ways Forward. Int J Disaster Risk Sci, 10:14–27.

Mamoon WB, N Jahan, F Abdullah et al. 2024. Modeling the Impact of Climate Change on Streamflow in the Meghna River Basin: An Analysis Using SWAT and CMIP6 Scenarios. Water, 16(8):1117.

Mancosu N, RL Snyder, G Kyriakakis et al. 2015. Water Scarcity and Future Challenges for Food Production. Water, 7(3):975–992.

Mancuso G, S Lavrnić, A Toscano 2020. Reclaimed Water to Face Agricultural Water Scarcity in the Mediterranean Area: An Over-View Using Sustainable Development Goals Preliminary Data, in: Wastewater Treat. Reuse—Present Futur. Perspect Technol Dev Manag, 5:113–143.

Mao G, J Liu, F Han et al. 2020. Assessing the Interlinkage of Green and Blue Water in an Arid Catchment in Northwest China. Environ Geochem Health, 42:933–95.

Marcantonio RA 2018. Water Insecurity, Illness and Other Factors of Everyday Life: A Case Study From Choma District, Southern Province, Zambia. Water SA, 44(4):653–663.

Margat J, JV Gun. 2013. Groundwater Around the World: A Geographic Synopsis (1st ed.). CRC Press: London. https://doi.org/10.1201/b13977

Margat J, J van der Gun. 2013. Groundwater Around the World. CRC Press/Balkema.

Marhaento H, MJ Booij, THM Rientjes et al. 2017. Attribution of Changes in the Water Balance of a Tropical Catchment to Land Use Change Using the SWAT model. Hydrol Process, 31:2029–2040.

Markstrom SL, RS Regan, LE Hay et al. 2015. PRMS-IV, the Precipitation-Runoff Modeling System; Version 4. U.S. Geological Survey, Reston, VA, USA. Techniques and Methods, 2015, Book 6, Chap. B7; p. 158.

Marselina M, F Wibowo, A Mushfiroh 2022. Water Quality Index Assessment Methods for Surface Water: A Case Study of the Citarum River in Indonesia. Heliyon, 8:e09848.

Martínez-Valderrama J, J Olcina, G Delacámara et al. 2023. Complex Policy Mixes are Needed to Cope with Agricultural Water Demands Under Climate Change. Water Resour Manage, 37:2805–2834. https://doi.org/10.1007/s11269-023-03481-5

Mashizha TM, HM Tirivangasi 2023. Climate Change, Food Security and Poverty. In: The Palgrave Handbook of Global Social Problems. Palgrave Macmillan, Cham https://doi.org/10.1007/978-3-030-68127-2_182 1.

Masia S, J Susnik, S Marras et al. 2018. Assessment of Irrigated Agriculture Vulnerability Under Climate Change in Southern Italy. Water, 10:209.

Masoud MHZ, JM Basahi, A Alqarawy et al 2024. Flash Flood Prediction in Southwest Saudi Arabia Using GIS Technique and Surface Water Models. Appl Water Sci, 14:61.

Mateescu E, M Smarandache, N Jeler et al. 2013. Drought Conditions and Management Strategies in Romania. Initiative on "Capacity Development to Support National Drought

Management Policy" (WMO, UNCCD, FAO and UNW-DPC) – Country Report, available at: http://www.droughtmanagement.info/literature/UNW-DPC_NDMP_Country_Report_Romania_2013.pdf

Mazvimavi D 2010. Investigating Changes over Time of Annual Rainfall in Zimbabwe. Hydrol Earth Syst Sci, 14:2671–2679.

Mbewu AD, DE Elephant, H Motsi et al. 2024. Climate Change Effects on Water Footprint of Crop Production: A Meta-Analysis. Environ Chall, 17:101033.

McClain ME 2013. Balancing Water Resources Development and Environmental Sustainability in Africa: A Review of Recent Research Findings and Applications. AMBIO, 42:549–565.

McDonnell JJ 2017. Beyond the Water Balance. Nature Geoscience, 11:496.

McElrone AJ, B Choat, GA Gambetta et al. 2013. Water Uptake and Transport in Vascular Plants. Nat. Educ. Knowledge, 4(5):6.

McKenna OP, OE Sala 2018. Groundwater Recharge in Desert Playas: Current Rates and Future Effects of Climate Change. Environ Res Lett, 13:014025.

McNamara I, M Flörke, T Uschan et al. 2024. Estimates of Irrigation Requirements Throughout Germany Under Varying Climatic Conditions. Agr Water Manage, 291:108641.

Meddi M, H Meddi, AA Assani 2009. Drought Assessment Using the Standardized Precipitation Index (SPI) in Algeria. Water Res Manag, 23:2711–2723.

Mehla MK 2022. Regional Water Footprint Assessment for a Semi-Arid Basin in India. Peer J, 10:e14207.

Mehta P, S Siebert, M Kummu et al. 2024. Half of Twenty-First Century Global Irrigation Expansion has been in Water-Stressed Regions. Nat Water, 2:254–261.

Mekonnen MM, W Gerbens-Leenes 2020. The Water Footprint of Global Food Production. Water, 12(10):2696. https://doi.org/10.3390/w12102696.

Mekonnen MM, AY Hoekstra 2012. A Global Assessment of the Water Footprint of Farm Animal Products. Ecosystems, 15:401–415. https://doi.org/10.1007/s10021-011-9517-8.

Mekonnen MM, AY Hoekstra 2016. Four Billion People Facing Severe Water Scarcity. Sci Adv, 2:e1500323.

Melching CS, CG Yoon 1996. Key sources of Uncertainty in QUAL2E Model of Passaic River. ASCE J Water Resour. Plan Manag, 122(2):105–113.

Mendez M, LA Calvo-Valverde, P Imbach et al. 2022. Hydrological Response of Tropical Catchments to Climate Change as Modeled by the GR2M Model: A Case Study in Costa Rica. Sustainability, 14(24):16938.

Meneguzzo F, F Zabini. 2021. Water Conservation and Resource Efficiency in Agriculture. In: Agri-Food and Forestry Sectors for Sustainable Development. Sustainable Development Goals Series. Springer, Cham. https://doi.org/10.1007/978-3-030-66284-4_6.

Mengel K, EA Kirkby, H Kosegarten et al. 2001. Plant Water Relationships. In: Mengel, K, EA Kirkby, H Kosegarten, T Appel (eds), Principles of Plant Nutrition. Springer, Dordrecht. https://doi.org/10.1007/978-94-010-1009-2_4.

Meran G, M Siehlow, C von Hirschhausen. 2021. Integrated Water Resource Management: Principles and Applications. In: The Economics of Water. Springer Water. Springer, Cham. https://doi.org/10.1007/978-3-030-48485-9_3.

Meran G, M Siehlow, C von Hirschhausen. 2021. Water Availability: A Hydrological View. In: The Economics of Water. Springer Water. Springer, Cham. https://doi.org/10.1007/978-3-030-48485-9_2.

Meuwissen MPM et al. 2019. A Framework to Assess the Resilience of Farming Systems. Agric Syst, 176:102656.

Meybeck M 1995. Global Distribution of Lakes. In: Lerman, A, DM Imboden, JR Gat (eds), Physics and Chemistry of Lakes. Springer, Berlin, Heidelberg. https://doi.org/10.1007/978-3-642-85132-2_1.

Meza I, E Eyshi Rezaei, S Siebert et al. 2021. Drought Risk for Agricultural Systems in South Africa: Drivers, Spatial Patterns, and Implications for Drought Risk Management. Sci Total Environ, 799:149505.

Mifsud Scicluna B, C Galdies 2025. Assessing the Impact of Temperature and Precipitation Trends of Climate Change on Agriculture Based on Multiple Global Circulation Model Projections in Malta. Big Data Cognitive Computing, 9(4):105.

Mila AJ, MH Ali, AR Akanda et al. 2017. Effects of Deficit Irrigation on Yield, Water Productivity and Economic Return of Sunflower. Cogent Food Agric, 3: 1287619.

Miller JD, CL Workman, SV Panchang et al. 2021. Water Security and Nutrition: Current Knowledge and Research Opportunities. Adv Nutr, 12(6):2525–2539. https://doi.org/ 10.1093/advances/nmab075.

Milly PCD, KA Dunne 2020. Colorado River Flow Dwindles as Warming-Driven Loss of Reflective Snow Energizes Evaporation. Science, 367:1252–1255.

Mirzabaev A, L Olsson, RB Kerr et al. 2023. Climate Change and Food Systems. In: von Braun, J, K Afsana, LO Fresco (eds), Science and Innovations for Food Systems Transformation. Springer, Cham. https://doi.org/10.1007/978-3-031-15703-5_27.

Mishra BK, P Kumar, C Saraswat et al. 2021. Water Security in a Changing Environment: Concept, Challenges and Solutions. Water, 13(4):490.

Misra AK 2014. Climate Change and Challenges of Water and Food Security. Int J Sustainable Built Environ, 3(1):153–165.

Mitsch WJ, L Zhang, D Marois et al. 2015. Protecting the Florida Everglades Wetlands with Wetlands: Can Stormwater Phosphorus be Reduced to Oligotrophic Conditions? Ecol Eng, 80:8–19.

Moges E, Y Demissie, L Larsen, F Yassin 2021. Review: Sources of Hydrological Model Uncertainties and Advances in Their Analysis. Water, 13(1):28.

Mohajerani H, DA Zema, ME Lucas, - Borja et al. 2021. Understanding the Water Balance and Its Estimation Methods. In: Rodrigo-Comino J (eds), Precipitation (pp. 193–221). Elsevier, Amsterdam, The Netherlands.

Mohammadi A, A Parvaresh Rizi, N Abbasi 2019. Field Measurement and Analysis of Water Losses at the Main and Tertiary Levels of Irrigation Canals: Varamin Irrigation Scheme, Iran. Glob Ecol Conserv, 18:e00646.

Molden D, T Oweis, P Steduto et al. 2010. Improving Agricultural Water Productivity: Between Optimism and Caution. Agric Water Manag, 97(4):528–535.

Molden D, M Vithanage, C de Fraiture et al. 2011. Water Availability and Its Use in Agriculture. Treatise Water Sci, 4:707–732.

Molle F, P Wester, P Hirsch, JR Jensen, H Murray-Rust, V Paranjpye, S Pollard, P van der Zaag. 2007. River Basin Development and Management. IWMI Books, Reports H040208, International Water Management Institute.

Mollel GR, DMM Mulungu, J Nobert et al. 2023. Assessment of Climate Change Impacts on Hydrological Processes in the Usangu Catchment of Tanzania under CMIP6 Scenarios. J Water Clim Change, 14:4162–4182.

Mondal P, M Walter, J Miller et al. 2023. The Spread and Cost of Saltwater Intrusion in the US Mid-Atlantic. Nat. Sustain, 6:1352–1362.

Montecelos-Zamora Y, T Cavazos, T Kretzschmar et al. 2018. Hydrological Modeling of Climate Change Impacts in a Tropical River Basin: A Case Study of the Cauto River, Cuba. Water, 10(9):1135.

Monteith JL 1965. Evaporation and Environment. In: Fogg, GE (ed), The State and Movement of Water in Living Organisms (pp. 205–234). Cambridge University Press: Cambridge.

Moore RJ 1985. The Probability-Distributed Principle and Runoff Production at Point and Basin Scales. Hydrol Sci J, 30(2):273–297.

Morell FJ, J Lampurlanes, J Alvaro-Fuentes et al. 2011. Yield and Water Use Efficiency of Barley in a Semiarid Mediterranean Agroecosystem: Long-Term Effects of Tillage and N Fertilization. Soil Till Res, 117:76–84.

Moriasi DN, JG Arnold, MW Van Liew et al. 2007. Model Evaluation Guidelines for Systematic Quantification of Accuracy in Watershed Simulations. Transactions of the ASABE, 50:885–900.

Morris M 1991. Factorial Sampling Plans for Preliminary Computational Experiments. Technometrics, 33(2):161–174.

Morton LW 2015. Achieving Water Security in Agriculture: The Human Factor. Agron J, 107(4):1557–1560.

Mosley LM 2015. Drought Impacts on The Water Quality of Freshwater Systems; Review and Integration. Earth-Sci Rev, 140:203–214.

Mostafa SM, O Wahed, WY El-Nashar et al. 2021. Impact of Climate Change on Water Resources and Crop Yield in the Middle Egypt Region. J Water Supply Res Technol-Aqua, 70(7):1066–1084.

Moyer JD, S Hedden 2020. Are We on the Right Path to Achieve the Sustainable Development Goals? World Dev, 127:104749.

Mpakairi K, T Dube, M Sibanda et al. 2024. Remote Sensing Crop Water Productivity and Water Use for Sustainable Agriculture During Extreme Weather Events in South Africa. Int J Appl Earth Obs Geoinf, 129:103833.

Mueller B, M Hauser, C Iles 2015. Lengthening of the Growing Season in Wheat and Maize Producing Regions. Weather Clim Extrem, 9:47–56.

Mugagga F, B Nabaasa 2016. The Centrality of Water Resources to the Realization of Sustainable Development Goals (SDG). A Review of Potentials and Constraints on the African continent. Int Soil Water Conserv Res, 4(3):215–223.

Mukherjee A, BR Scanlon, A Aureli, et al. 2021. Global Groundwater: Source, Scarcity, Sustainability, Security, and Solutions. Elsevier, Amsterdam, Netherlands. https://doi.org/10.1016/C2018-0-03156-4.

Mupaso N, G Makombe, R Mugandani 2023. Smallholder Irrigation and Poverty Reduction in Developing Countries: A Review. Heliyon, 9(2):e13341.

Murad A 2010. An Overview of Conventional and Non-Conventional Water Resources in Arid Region: Assessment and Constrains of the United Arab Emirates (UAE). J Water Resource Prot, 2:181–190.

Musonda B, Y Jing, V Iyakaremye et al. 2020. Analysis of Long-Term Variations of Drought Characteristics Using Standardized Precipitation Index over Zambia. Atmosphere, 11(12):1268.

Musse S 2021. Exploring the Cornerstone Factors that Cause Water Scarcity in Some Parts of Africa, Possible Adaptation Strategies and Quest for Food Security. Agricultural Sciences, 12:700–712.

Mutharika BW. 2010. The African Food Basket: Innovations, Interventions, and Strategic Partnerships. Keynote address of the President of the Republic of Malawi and Chairperson for the African Union at the Lecture and Video Conference at Boston University, October 1, 2010.

Mwadzingeni L, R Mugandani, P Mafongoya 2022. Risks of Climate Change on Future Water Supply in Smallholder Irrigation Schemes in Zimbabwe. Water, 14(11):1682.

Myers DT, DL Ficklin, SM Robeson 2023. Hydrologic Implications of Projected Changes in Rain-On-Snow Melt for Great Lakes Basin Watersheds. Hydrol Earth Syst Sci, 27:1755–1770.

Myers SS, A Zanobetti, I Kloog et al. 2014. Increasing CO_2 Threatens Human Nutrition. Nature, 510:139–142.

Nairizi S. 2017. Irrigated Agriculture Development under Drought and Water Scarcity. Int Commission Irrigation Drainage, 1–168.

Narsimlu B, AK Gosain, BR Chahar 2013. Assessment of Future Climate Change Impacts on Water Resources of Upper Sind River Basin, India Using SWAT Model. Water Resour Manage, 27:3647–3662.

Nasonova ON, YM Gusev, EM Volodin et al. 2018. Application of the Land Surface Model SWAP and Global Climate Model INMCM4.0 for Projecting Runoff of Northern Russian Rivers. 2. Projections and Their Uncertainties. Water Resour, 45(2):S85–S92.

Nava LF, S Sandoval, - Solis 2015. A lock-in Transboundary Water Management Regime: the case of the Rio Grande/Bravo Basin. In: World Water Congress, 25–29 May 2015, Edinburgh, Scotland.

Negm AM. 2018. Groundwater in the Nile Delta. The Handbook of Environmental Chemistry. Springer, Cham. https://doi.org/10.1007/978-3-319-94283-4.

Ngo TML, SJ Wang, PY Chen 2024. Assessment of Future Climate Change Impacts on Groundwater Recharge Using Hydrological Modeling in the Choushui River Alluvial Fan, Taiwan. Water, 16(3):419.

Nguyen H, A Thompson, C Costello 2023. Impacts of Historical Droughts on Maize and Soybean Production in the Southeastern United States. Agric Water Manag, 281:108237. https://doi.org/10.1016/j.agwat.2023.108237.

Nijssen B, G O'Donnell, A Hamlet et al. 2001. Hydrologic Sensitivity of Global Rivers to Climate Change. Clim. Change, 50(1–2):143–175.

Niraula R, T Meixner, F Dominguez et al. 2017. How Might Recharge Change under Projected Climate Change in the Western U.S.? Geophys Res Lett, 44(20):407–10, 418.

Nistor MM 2019. Climate Change Effect on Groundwater Resources in Southeast Europe During 21st Century. Quat. Int, 504:171–180.

Niu L, W Qin, Y You et al. 2023. Effects of Precipitation Variability and Conservation Tillage on Soil Moisture, Yield and Quality Of Silage Maize. Front Sustain Food Syst, 7:1198649.

Niyigaba E, A Twizerimana, I Mugenzi et al. 2019. Winter Wheat Grain Quality, Zinc and Iron Concentration Affected by a Combined Foliar Spray of Zinc and Iron Fertilizers. Agronomy, 9(5):250. https://doi.org/10.3390/agronomy9050250.

Nkomozepi T, S Chung 2014. The Effect of Climate Change on the Water Resources of the Geumho River Basin, Republic of Korea. J Hydro-Environ Res, 8(4):358–366.

Nor Diana MI, NA Zulkepli, C Siwar et al. 2022. Farmers' Adaptation Strategies to Climate Change in Southeast Asia: A Systematic Literature Review. Sustainability, 14:3639.

Nouri H, B Stokvis, S Chavoshi Borujeni et al. 2020. Reduce Blue Water Scarcity and Increase Nutritional and Economic Water Productivity Through Changing the Cropping Pattern in a Catchment. J. Hydrol, 588:125086.

Novoa V, C Rojas, O Rojas et al. 2024. A Temporal Analysis of the Consequences of the Drought Regime on the Water Footprint of Agriculture in the Guadalupe Valley, Mexico. Sci Rep, 14:6114.

Nunez JA, S Aguiar, EG Jobbagy et al. 2024. Climate Change and Land Cover Effects on Water Yield in a Subtropical Watershed Spanning the Yungas-Chaco Transition of Argentina. J Environ Manag, 358:120808.

Nury AH, A Sharma, R Mehrotra et al. 2022. Projected Changes in the Tibetan Plateau Snowpack Resulting from Rising Global Temperatures. J Geophys Res Atmos, 127: e2021JD036201.

Obuobie E, K Kankam-Yeboah, B Amisigo et al. 2012. Assessment of Vulnerability of River Basins in Ghana to Water Stress Conditions Under Climate Change. J Water Clim Change, 3(4):276–286.

Ochoa-Noriega CA, JA Aznar-Sánchez, JF Velasco-Muñoz et al. 2020. The Use of Water in Agriculture in Mexico and Its Sustainable Management: A Bibliometric Review. Agronomy, 10(12):1957.

Ochoa–Tocachi BF, J Cuadros, – Adriazola, E Arapa et al. 2022. Guide for Hydrologic Modeling of Natural Infrastructure. Forest Trends, Lima, Peru.

Ochwo OCM, G Okwir, JR Selemani et al. 2025. Groundwater Recharge Assessment Under Climate Change Scenarios: A Case Study of Kiryandongo, Uganda. J. Water Clim. Change, 16(5):1877–1894.

Oduor BO, MA Campo-Bescós, N Lana-Renault et al. 2023. Effects of Climate Change on Streamflow and Nitrate Pollution in An Agricultural Mediterranean Watershed in Northern Spain. Agric. Water Manag, 285:108378.

OECD 2014. Climate Change, Water and Agriculture: Towards Resilient Systems, OECD Studies onWater, OECD Publishing.

Ohba M, R Arai, T Sato et al. 2025. Projected Climate Change Impacts on Hydrological Droughts in Japan: Dependency on Climate And Weather Patterns. Clim Dyn, 63:119.

Ojima DS, RT Conant, WJ Parton et al. 2021. Recent Climate Changes across the Great Plains and Implications for Natural Resource Management Practices. Rangel Ecol Manag, 78:180–190.

Okutan P, A Akkoyunlu 2021. Identification of Water Use Behavior and Calculation of Water Footprint: A Case Study. Appl Water Sci, 11:127. https://doi.org/10.1007/s13201-021-01459-5.

Orkodjo TP, G Kranjac-Berisavijevic, FK Abagale 2022. Impact of Climate Change on Future Precipitation Amounts, Seasonal Distribution, and Streamflow in the Omo-Gibe Basin, Ethiopia. Heliyon, 8(6):e09711.

Ortolano L, KK Cushing 2002. Grand Coulee Dam 70 Years Later: What Can We Learn? Int J Water Resour Dev, 18(3):373–390.

Ouda S, AEH Zohry 2020. Water Scarcity Leads to Food Insecurity. In: Deficit Irrigation. Springer, Cham. https://doi.org/10.1007/978-3-030-35586-9_1.

Owusu S, O Cofie, M Mul 2022. The Significance of Small Reservoirs in Sustaining Agricultural Landscapes in Dry Areas of West Africa: A Review. Water, 14(9):1440.

Özdoğan-Sarıkoç G, F Dadaser-Celik 2024. Physically based vs. Data-driven Models for Streamflow and Reservoir Volume Prediction at a Data-Scarce Semi-Arid Basin. Environ Sci Pollut Res, 31:39098–39119.

Pabuayon ILB, S Singh, GL Ritchie 2019. Effects of Deficit Irrigation on Yield and Oil Content of Sesame, Safflower, and Sunflower. Agron. J, 111:3091–3098.

Parandvash GH, H Chang 2016. Analysis of Long-Term Climate Change on Per Capita Water Demand in Urban Versus Suburban Areas in the Portland Metropolitan Area, USA. J Hydrol, 538:574–586.

Parra-López C, S Ben Abdallah, G Garcia-Garcia 2025. Digital Technologies for Water Use and Management in Agriculture: Recent Applications and Future Outlook. Agric Water Manag, 309:109347.

Patanè C, S Tringali, O Sortino 2011. Effects of Deficit Irrigation on Biomass, Yield, Water Productivity and Fruit Quality of Processing Tomato under Semi-Arid Mediterranean Climate Conditions. Sci Hortic, 129:590–596.

Payus C, L Ann Huey, F Adnan et al. 2020. Impact of Extreme Drought Climate on Water Security in North Borneo: Case Study of Sabah. Water, 12(4):1135.

Pereira LS, T Oweis, A Zairi 2002. Irrigation Management Under Water Scarcity. Agric. Water Manag, 57:175–206.

Pereira LS, P Paredes, N Jovanovic et al. 2020. Soil Water Balance Models for Determining Crop Water and Irrigation Requirements and Irrigation Scheduling Focusing on the FAO56 Method and the Dual K_c Approach. Agric Water Manag, 241:106357.

Peres DJ, R Modica, A Cancelliere 2019. Assessing Future Impacts of Climate Change on Water Supply System Performance: Application to the Pozzillo Reservoir in Sicily Italy. Water, 11:2531.

Perkins-Kirkpatrick SE, PB Gibson 2017. Changes in Regional Heatwave Characteristics as a Function of Increasing Global Temperature. Sci. Rep, 7:12256.

Petersen L, M Heynen, F Pellicciotti 2019. Freshwater Resources: Past, Present, Future. https://doi.org/10.1002/9781118786352.wbieg0712.pub2.

Petersen-Perlman JD, I Aguilar-Barajas, SB Megdal 2022. Drought and Groundwater Management: Interconnections, Challenges, and Policy Responses. Curr Opin Environ Sci Health, 28:100364.

Phillips CL, N Nickerson. 2015. Soil Respiration. Reference Module in Earth Systems and Environmental Sciences. Elsevier: Amsterdam.

Pimentel D, Berger B, Filiberto D, et al. 2004. Water Resources: Agricultural and Environmental Issues. *BioScience*, 54:909–918.

Piniewski M, CL Laizé, MC Acreman et al. 2014. Effect of Climate Change on Environmental Flow Indicators in the Narew Basin, Poland. J Environ Qual, 43(1):155–67.

Pisinaras V, C Petalas, GD Gikas et al. 2010. Hydrological and Water Quality Modeling in a Medium-Sized Basin using the Soil and Water Assessment Tool (SWAT). Desalination, 250:274–286.

Pixley KV, JE Cairns, S Lopez-Ridaura et al. 2023. Redesigning Crop Varieties to Win the Race between Climate Change and Food Security. Mol Plant, 16(10):1590–1611.

Pokhrel BK, KP Paudel, E Segarra 2018. Factors Affecting the Choice, Intensity, and Allocation of Irrigation Technologies by U.S. Cotton Farmers. Water, 10(6):706.

Poore J, T Nemecek 2018. Reducing Food's Environmental Impacts Through Producers and Consumers. Science, 360:987–992.

Postel SL, GC Daily, PR Ehrlich 1996. Human Appropriation of Renewable Fresh Water. Science, 271:785–788.

Pritchard HD 2019. Asia's Shrinking Glaciers Protect Large Populations from Drought Stress. Nature, 569:649–654.

Prosser IP, FHS Chiew, M Stafford Smith 2021. Adapting Water Management to Climate Change in the Murray–Darling Basin, Australia. Water, 13(18):2504.

Pulido-Esquivel AY, JV Prado-Hernández, JC Buendía-Espinoza et al. 2025. Agroforestry Systems Enhance Soil Moisture Retention and Aquifer Recharge in a Semi-Arid Mexican Valley. Water, 17(10):1488.

Qiao Y, H Zhang, B Dong et al. 2010. Effects of Elevated CO_2 Concentration on Growth and Water Use Efficiency of Winter Wheat Under Two Soil Water Regimes. Agric Water Manag, 97(11):1742–1748.

Qin G, B Wu, X Dong et al. 2023. Evolution of Groundwater Recharge-Discharge Balance in the Turpan Basin of China During 1959–2021. J Arid Land, 15:1037–1051.

Qin Y, ND Mueller, S Siebert et al. 2019. Flexibility and Intensity of Global Water Use. Nat Sustain, 2:515–523.

Qiu H, T Zhou, L Zou et al. 2024. Future Changes in Precipitation and Water Availability Over the Tibetan Plateau Projected by Cmip6 Models Constrained by Climate Sensitivity. Atmos Ocean Sci Lett, 17:100537.

Querner E, J den Besten, R van Veen et al. 2022. A Scenario Analysis of Climate Change and Adaptation Measures to Inform Dutch policy in the Netherlands. J Water Land Dev, 54:177–183.

Qureshi ME, MA Hanjra, J Ward 2013. Impact of Water Scarcity in Australia on Global Food Security in an Era of Climate Change. Food Policy, 38:136–145.

Rajabi AM, S Alizadehnia, A Sohrabi Bidar 2025. A Review of Studies on Underground Dam Site Selection. Phys Chem Earth (Pt AB,C), 140:103995.

Rajosoa AS, C Abdelbaki, KA Mourad 2022. Assessing the Impact of Climate Change on The Medjerda River Basin. Arab. J. Geosci, 15:1052.

Rameshwaran P, VA Bell, NH Davies et al. 2021. How Might Climate Change Affect River Flows Across West Africa? Clim Change, 169:21.

Ramírez-Cuesta JM, JM Mirás-Avalos, JS Rubio-Asensio et al. 2018. A Novel ArcGIS Toolbox for Estimating Crop Water Demands By Integrating the Dual Crop Coefficient Approach With Multi-Satellite Imagery. Water, 11(1):38.

Rana R, R Ganguly, AK Gupta 2018. Indexing Method for Assessment of Pollution Potential of Leachate from Non-Engineered Landfill Sites and Its Effect on Groundwater Quality. Environ Monit Assess, 190:46.

Ray DK, PC West, M Clark et al. 2019. Climate Change has Likely Already Affected Global Food Production. PLoS One, 14(5):e0217148.

Ray RL, A Fares, E Risch 2018. Effects of Drought on Crop Production and Cropping Areas in Texas. Agric Environ Lett, 3:170037.

Reardon T, R Echeverria, J Berdegue et al. 2019. Rapid Transformation of Food Systems in Developing Regions: Highlighting the Role of Agricultural Research and Innovations. Agric Syst, 172:47–59.

Refsgaard JC, B Storm 1990. Construction, Calibration and Validation of Hydrological Models. In: Abbott MB, JC Refsgaard (eds) Distributed Hydrological Modelling. Water Science and Technology Library, vol 22. Springer, Dordrecht.

Renault D, R Wahaj, S Smits, 2013. Multiple Uses of Water Services in Large Irrigation System: The MASSMUS Approach. FAO Irrigation and Drainage Paper 67, FAO, Rome, Italy. http://www.fao.org/docrep/018/i3414e/i3414e.pdf.

Revenga C, T Tyrrell 2018. Major River Basins of the World. In: Finlayson, C, G Milton, R Prentice, N Davidson (eds), The Wetland Book. Springer, Dordrecht. https://doi.org/10.1007/978-94-007-4001-3_211.

Rezaei KS, A Sharaai, L Manaf et al. 2019. Assessing Ground and Surface Water Scarcity Indices Using Ground and Surface Water Footprints in the Tehran Province of Iran. Appl. Ecol. Environ. Res, 17:4985–4997.

Ribeiro L. 2007. Groundwater in the Southern Member States of the European Union: an Assessment of Current Knowledge and Future Prospects, European Academies' Science Advisory Council. Germany.

Richey AS, BF Thomas, MH Lo et al. 2015. Quantifying Renewable Groundwater Stress With GRACE. Water Resour Res, 51:5217–5238. https://doi.org/10.1002/2015WR017349.

Richter BD 2025. Betting the Farm While Irrigation Supplies Dwindle. Discov Agric, 3:24.

Riediger J, B Breckling, RS Nuske et al. 2014. Will Climate Change Increase Irrigation Requirements in Agriculture of Central Europe? A Simulation Study for Northern Germany. Environ Sci Eur, 26:18.

Ritchie H, M Roser. (2018) Water Use and Stress. United Kingdom: Our World in Data.org.

Rivas I, A Güitrón, M Montero 2011. Hydrologic Vulnerability to Climate Change of The Lerma-Chapala Basin, Mexico. Sustainability Today, 167:297–308.

Rivera A, T Gunda, GM Hornberger 2018. Minimizing Irrigation Water Demand: An Evaluation of Shifting Planting Dates in Sri Lanka. Ambio, 47(4):466–476.

Rivera JA, S Otta, C Lauro et al. 2021. A Decade of Hydrological Drought in Central-Western Argentina. Front Water, 3:640544.

Rockström J, M Falkenmark, L Karlberg et al. 2009. Future Water Availability for Global Food Production: The Potential of Green Water for Increasing Resilience to Global Change. Water Resour Res, 45:W00A12.

Rockström J, L Karlberg, SP Wani et al. 2010. Managing Water in Rainfed Agriculture—The need for a Paradigm Shift. Agric Water Manag, 97:543–550.

Roetter RP, H Van Keulen 2007. Food Security. In: Roetter, RP, H Van Keulen, M Kuiper, J Verhagen, HH Van Laar (eds), Science for Agriculture and Rural Development in Low-Income Countries. Springer, Dordrecht. https://doi.org/10.1007/978-1-4020-6617-7_3.

Rogers, J, T Borisova, J Ullman et al. 2014. Factors Affecting the Choice of Irrigation Systems for Florida Tomato Production. UF-IFAS Extension FE960.

Rojas-Downing MM, AP Nejadhashemi, T Harrigan et al. 2017. Climate Change and Livestock: Impacts, Adaptation, and Mitigation. Clim Risk Manag, 16:145–163.

Rondhi M, SJH Suherman, CB Hensie et al. 2024. Urbanization Impacts on Rice Farming Technical Efficiency: A Comparison of Irrigated and Non-Irrigated Areas in Indonesia. Water, 16(5):651.

Rosa L, DD Chiarelli, MC Rulli et al. 2020. Global Agricultural Economic Water Scarcity. Sci Adv, 6(18):eaaz6031.

Rosa L, DD Chiarelli, C Tu et al. 2019. Global Unsustainable Virtual Water Flows in Agricultural Trade. Environ Res Lett, 14:114001.

Rosa L, M Sangiorgio 2025. Global Water Gaps Under Future Warming Levels. Nat Commun, 16(2025):1192. https://doi.org/10.1038/s41467-025-56517-2.

Rosegrant MW 1997. Water Resources in the Twenty-First Century: Challenges and Implications for Action. Food, Agriculture and the Environment. Discussion Paper 20, IFPRI, Washington.

Rost S, D Gerten, A Bondeau et al. 2008. Agricultural Green and Blue Water Consumption and Its Influence on the Global Water System. Water Resour Res, 44.

Roth G, G Harris, M Gillies et al. 2013. Water-Use Efficiency and Productivity Trends in Australian Irrigated Cotton: A Review. Crop Pasture Sci, 64(11–12):1033–1048.

Rufino PR, B Gücker, M Volk et al. 2025. Modeling the Nexus of Climate Change and Deforestation: Implications for the Blue Water Resources of the Jari River, Amazonia. Water, 17(5):660.

Rusli SR, VF Bense, SMT Mustafa et al. 2017. The Impact of Future Changes in Climate Variables and Groundwater Abstraction on Basin-Scale Groundwater Availability. Hydrol Earth Syst Sci, 28:5107–5131.

Russo T, K Alfredo, J Fisher 2014. Sustainable Water Management in Urban, Agricultural, and Natural Systems. Water, 6:3934–3956.

Sadoff CW, D Grey 2005. Cooperation on International Rivers: A Continuum for Securing and Sharing Benefits. Water Int, 30:420–427.

Sadok W, JR Lopez, KP Smith 2021. Transpiration Increases Under High-Temperature Stress: Potential Mechanisms, Trade-Offs, and Prospects for Crop Resilience in a Warming World. Plant Cell Environ, 44(7):2102–2116.

Sadyrov S, E Isaev, K Tanaka et al. 2025. High-Resolution Assessment of Climate Change Impacts on the Surface Energy and Water Balance in the Glaciated Naryn River Basin, Central Asia. J Environ Manage, 374:124021.

Saha D, S Sahu 2015. A Decade of Investigations on Groundwater Arsenic Contamination in Middle Ganga Plain, India. Environ Geochem Health, 38:315–337.

Saha PP, K Zeleke 2015. Rainfall-Runoff Modelling for Sustainable Water Resources Management: SWAT Model Review in Australia. In: Setegn, S, M Donoso (eds), Sustainability of Integrated Water Resources Management. Springer, Cham. https://doi.org/10.1007/978-3-319-12194-9_29.

Saha S, S Moorthi, H Pan et al. 2010. The NCEP Climate Forecast System Reanalysis. Bull Am Meteorol Soc, 91(8):1015–1058.

Sahu C, SK Chandniha 2025. A Brief Review on Hydrological Modelling. Int J Environ Clim Change, 15:352–368.

Sahu MK, HR Shwetha, GS Dwarakish 2023. State-of-the-Art Hydrological Models and Application of the HEC-HMS Model: A Review. Model Earth Syst Environ, 9:3029–3051.

Salehin M et al. 2018. Mechanisms and Drivers of Soil Salinity in Coastal Bangladesh. In: Nicholls, R, C Hutton, W Adger, S Hanson, M Rahman, M Salehin (eds), Ecosystem Services for Well-Being in Deltas. Palgrave Macmillan, Cham. https://doi.org/10.1007/978-3-319-71093-8_18.

Sambo P, C Nicoletto, A Giro et al. 2019. Hydroponic Solutions for Soilless Production Systems: Issues and Opportunities in a Smart Agriculture Perspective. Front Plant Sci, 10:923.

Sand-Jensen K. 2013. Freshwater Ecosystems, Human Impact On (pp. 570–586). Academic Press, Cambridge. https://doi.org/10.1016/B978-0-12-384719-5.00369-5.

Santos L, J Andersson, B Arheimer 2022. Evaluation of Parameter Sensitivity of a Rainfall-Runoff Model Over a Global Catchment Set. Hydrolog Sci J, 67:342–357.

Sauchyn D, D Davidson, M Johnston 2020. Prairie Provinces; Chapter 4 in Canada in a Changing Climate: Regional Perspectives Report, In: F.J. Warren, N. Lulham and D.S. Lemmen (eds.). Government of Canada, Ottawa, Ontario.

Savenije HHG 2000. Water Scarcity Indicators; The Deception of the Numbers. Phys Chem Earth Part B Hydrol Oceans Atmos, 25:199–204.

Schade-Poole K, G Möller 2016. Impact and Mitigation of Nutrient Pollution and Overland Water Flow Change on the Florida Everglades, USA. Sustainability, 8(9):940.

Schillinger J, G Özerol, S Güven-Griemert et al. 2020. Water in War: Understanding the Impacts of Armed Conflict on Water Resources and Their Management. WIREs Water, 7(6):e1480.

Schmitt RJP, L Rosa 2024. Dams for Hydropower and Irrigation: Trends, Challenges, and Alternatives. Renewable and Sustainable Energy Reviews, 199:114439.

Schneider C, CLR Laizé, MC Acreman et al. 2013. How Will Climate Change Modify River Flow Regimes in Europe? Hydrol Earth Syst Sci, 17:325–339.

Schymanski SJ, O Or 2016. Wind Increases Leaf Water Use Efficiency. Plant Cell Environ, 39(7):1448–1459.

Schyns JF, AY Hoekstra, MJ Booij 2015. Review and Classification of Indicators of Green Water Availability and Scarcity. Hydrol Earth Syst Sci, 19:4581–4608.

Seck PA, A Diagne, S Mohanty et al. 2012. Crops that Feed the World 7: Rice. Food Sec, 4:7–24.

Seckler D, U Amarasinghe, DJ Molden, et al. 1998. World Water Demand and Supply, 1990 to 2025: Scenarios and Issues. IWMI, Colombo, Sri Lanka.

Seijger C, P Hellegers 2023. How Do Societies Reform their Agricultural Water Management Towards New Priorities for Water, Agriculture, and The Environment? Agric Water Manag, 277:108104.

Semba RD, S Askari, S Gibson et al. 2022. The Potential Impact of Climate Change on the Micronutrient-Rich Food Supply. Adv Nutr, 13(1):80–100.

Şen Z 2021. Reservoirs for Water Supply under Climate Change Impact—A Review. Water Resour Manage, 35:3827–3843.

Senent-Aparicio J, L Peñafiel, FJ Alcalá et al. 2024. Climate Change Impacts on Renewable Groundwater Resources in the Andosol-Dominated Andean Highlands. Ecuador. Catena, 236:107766.

Serrano A, I Cazcarro, M Martín-Retortillo et al. 2024. Europe's Orchard: The Role of Irrigation on the Spanish Agricultural Production. J Rural Stud, 110:103376.

Shahid S 2011. Impact of Climate Change on Irrigation Water Demand of Dry Season Boro Rice in Northwest Bangladesh. Climatic Change, 105:433–453.

Shandilya RN, E Bresciani, PK Kang et al. 2022. Influence of Hydrogeological and Operational Parameters on Well Pumping Capacity. J Hydrol, 608:127643.

Sharaunga S, M Mudhara 2018. Determinants of Farmers' Participation in Collective Maintenance of Irrigation Infrastructure in KwaZulu-Natal. Phys Chem Earth, 105:265–273.

Shiferaw N, L Habte, M Waleed 2025. Land use Dynamics and their Impact on Hydrology and Water Quality of A River Catchment: A Comprehensive Analysis and Future Scenario. Environ Sci Pollut Res, 32:4124–4136.

Shiklomanov IA 2004. World Water Balance. Encyclopedia of Life Support Systems UNESCO 2004.

Shiklomanov IA, J Rodda. 2003. World Water Resources at the Beginning of the Twenty-First Century. UNESCO, Cambridge UK.

Shiklomanov IA, JC Rodda 2003. World Water Resources at the Beginning of the 21st Century, International Hydrology Series, Cambridge University Press, Cambridge, UK.

Shit PK, PP Adhikary, B Bera et al. 2024. Resilient and Sustainable Water Management in Agriculture. Environ Sci Pollut Res, 31:54020–54025.

Shrestha A, AT Simmons, G Roth 2023. Water Use in Australian Irrigated Agriculture—Sentiments of Twitter Users. Water, 15(15):271.

Siddik MS, SS Tulip, A Rahman et al. 2022. The Impact of Land Use and Land Cover Change on Groundwater Recharge in Northwestern Bangladesh. J Environ Manag, 315:115130.

Silva JRM, MC de Oliveira Celeri, AC Borges et al. 2023. Greywater as a Water Resource in Agriculture: The Acceptance and Perception from Brazilian Agricultural Technicians. Agric Water Manag, 280:108227.

Singh N, A Pandey, V Singh 2024. Impact of Institutional Overlapping on Water Governance of Himalayan City, Nainital, India. Water Policy, 26(6):601–617.

Sîrodoev I, R Corobov, G Sîrodoev et al. 2022. Modelling Runoff within a Small River Basin under the Changing Climate: A Case Study of Using SWAT in the Bălțata River Basin (The Republic of Moldova). Land, 11(2):167.

Sismani G, V Pisinaras, G Arampatzis 2024. Water Governance for Climate-Resilient Agriculture in Mediterranean Countries. Water, 16(8):1103.

Sivakumar MVK, AC Ruane, J Camacho 2013: Climate Change in the West Asia and North Africa Region. In: M.V.K. Sivakumar, R. Lal, R. Selvaraju, and I. Hamdan (eds), Climate Change and Food Security in West Asia and North Africa (pp. 3–26). Springer, Netherlands. https://doi.org/10.1007/978-94-007-6751-5_1.

Sobaga A, F Habets, N Beaudoin et al. 2024. Decreasing Trend of Groundwater Recharge with Limited Impact Of Intense Precipitation: Evidence From Long-Term Lysimeter Data. J Hydrol, 637:131340.

Sobkowiak L, D Wrzesiński 2024. Impacts of Climate Change on Water Resources: Assessment and Modeling—First Edition. Water, 16:3578.

Sobol I 2001. Global Sensitivity Indices for Nonlinear Mathematical Models and Their Monte Carlo Estimates. Math. Comput Simul, 55(1–3):271–280.

Sommaruga R 2015. When Glaciers and Ice Sheets Melt: Consequences for Planktonic Organisms. J Plankton Res, 37(3):509–518.

Song JH, Y Her, X Yu et al. 2022a. Effect of Information-Driven Irrigation Scheduling on Water Use Efficiency, Nutrient Leaching, Greenhouse Gas Emission, and Plant Growth in South Florida. Agric Ecosyst Environ, 333:107954.

Song P, C Wang, G Ding et al. 2022b. Evaluating the Impact of Climate Change on Surface Water Resources in the Upper Ganjiang River Basin, China. J Water Clim Change, 13(3):1462–1476.

Sood A, L Muthuwatta, M McCartney 2013. A SWAT Evaluation of the Effect of Climate Change on the Hydrology of the Volta River Basin. Water Int, 38(3):297–311.

Soomro S, GS Solangi, AA Siyal et al. 2023. Estimation of Irrigation Water Requirement and Irrigation Scheduling for Major Crops Using the CROPWAT Model and Climatic Data. Water Pract Technol, 18:685–700.

Sordo-Ward A, I Granados, A Iglesias et al. 2019. Blue Water in Europe: Estimates of Current and Future Availability and Analysis of Uncertainty. Water, 11(3):420.

Souza SA, LN Rodrigues 2022. Increased Profitability and Energy Savings Potential With the Use of Precision Irrigation. Agric Water Manag, 270:107730.

Sowter A, MBC Amat, F Cigna et al. 2016. Mexico City Land Subsidence in 2014–2015 with Sentinel-1 IW Tops: Results Using the Intermittent SBAS (ISBAS) Technique. Int J Appl Earth Obs Geoinf, 52:230–242.

Steduto P, TC Hsiao, E Fereres, et al. 2012. Crop Yield Response to Water (p. 66). FAO, Rome, Italy.

Stefanova A, C Hesse, V Krysanova et al. 2019. Assessment of Socio-Economic and Climate Change Impacts on Water Resources in Four European Lagoon Catchments. Environ Manag, 64:701–720.

Steinfeld H, P Gerber, T Wassenaar et al. 2006. Livestock's Long Shadow: Environmental Issues and Options. Food and Agriculture Organization of the United Nations (FAO), Rome.

Steppe K 2018. The Potential of the Tree Water Potential. Tree Physiol, 38:937–940.

Stringer LC, A Mirzabaev 2021. Climate Change Impacts on Water Security in Global Drylands. One Earth, 4:851–864.

Stringer LC, A Mirzabaev, TA Benjaminsen et al. 2021. Climate Change Impacts on Water Security in Global Drylands. One Earth, 4:851–64.

Su YY, YH Weng, YW Chiu 2009. Climate Change and Food Security in East Asia. Asia Pac J Clin Nutr, 18(4):674–8.

Sucozhañay A, J Pesántez, R Guerrero-Coronel et al. 2024. Rainwater Harvesting as a Sustainable Solution for the Production of Urban Hydroponic Crops. Water Reuse, 14:177–89.

Sultan B, M Gaetani 2016. Agriculture in West Africa in the Twenty-First Century: Climate Change and Impacts Scenarios, and Potential for Adaptation. Front Plant Sci, 7:1262.

Summers WK 1972. Specific Capacities of Wells in Crystalline Rocks. Groundwater, 10(6): 37–47.

Sun J, G Zhang, Y Wu et al. 2024. Risk Assessment of Agricultural Green Water Security in Northeast China Under Climate Change. Sci China Earth Sci, 67:2178–2194.

Sun Y, W Dendi, DE Kim et al. 2019. Deriving Intensity–Duration–Frequency (IDF) Curves Using Downscaled in Situ Rainfall Assimilated with Remote Sensing Data. Geosci Lett, 6:17.

Sun Y, Z Nan, W Yang et al. 2023. Projecting China's Future Water Footprints and Water Scarcity Under Socioeconomic and Climate Change Pathways Using an Integrated Simulation Approach. Clim Serv, 30:100385.

Sutanto SJ, SB Zarzoza Mora, I Supit et al. 2024. Compound and Cascading Droughts and Heatwaves Decrease Maize Yields By Nearly Half In Sinaloa, Mexico. npj Nat Hazards, 1:26.

Sutcliffe JV 2004. Hydrology: A Question of Balance. IAHS Press.

Sylla MB, JS Pal, A Faye et al. 2018. Climate Change to Severely Impact West African Basin Scale Irrigation In 2 °C And 1.5 °C Global Warming Scenarios. Sci Rep, 8:14395.

Taka M, L Ahopelto, A Fallon et al. 2021. The Potential of Water Security in Leveraging Agenda 2030. One Earth, 4(2):258–268.

Tang J, H Bai, X Zhang et al. 2022. Reducing Potato Water Footprint by Adjusting Planting Date in the Agro-Pastoral Ecotone in North China. Ecol Modell, 474:110155.

Tarekegn N, B Abate, A Muluneh et al. 2022. Modeling the Impact of Climate Change on the Hydrology of Andasa Watershed. Model Earth Syst Environ, 8:103–119.

Tariq M, AN Rohith, R Cibin et al. 2024. Understanding Future Hydrologic Challenges: Modelling the Impact of Climate Change on River Runoff in Central Italy. Environ Chall, 15:100899.

Taub D 2010. Effects of Rising Atmospheric Concentrations of Carbon Dioxide on Plants. Nat Educ Knowl, 3(10):21.

Tauro F, J Selker, N van de Giesen et al. 2018. Measurements and Observations in the XXI Century (MOXXI): Innovation and Multi-Disciplinarity to Sense the Hydrological Cycle. Hydrol Sci J, 63(2):169–196.

Teferi E, T Kassawmar, W Bewket et al. 2025. Rainfed Agriculture in Ethiopia: A Systematic Review of Green Water Management Pathways to Improve Water and Food Security. Front Agron, 7:1418024.

Tenagashaw DY, TG Andualem 2022. Analysis and Characterization of Hydrological Drought Under Future Climate Change Using the SWAT Model in Tana Sub-basin, Ethiopia. Water Conserv Sci Eng, 7:131–142.

Terêncio DPS, LF Fernades, RM Cortes et al. 2020. Flood Risk Attenuation in Critical Zones of Continental Portugal Using Sustainable Detention Basins. Sci Total Environ, 721:137727.

Terrer C, RP Phillips, BA Hungate et al. 2021. A Trade-Off Between Plant and Soil Carbon Storage Under Elevated Co2. Nature, 591:599–603.

Teshome FT, HK Bayabil, B Schaffer et al. 2023. Exploring Deficit Irrigation as a Water Conservation Strategy: Insights from Field Experiments and Model Simulation. Agric Water Manag, 289:108490.

Teweldebrihan MD, MO Dinka 2025. Sustainable Water Management Practices in Agriculture: The Case of East Africa. Encyclopedia, 5(1):7.

Tezzo X, SR Bush, P Oosterveer et al. 2021. Food System Perspective on Fisheries and Aquaculture Development in Asia. Agric Hum, 38:73–90.

Thenkabail PS, MA Hanjra, V Dheeravath et al. 2011. Global Croplands and Their Water Use from Remote Sensing and Nonremote Sensing Perspectives. In Weng, Q. (ed.), Advances in Environmental Remote Sensing-Sensors, Algorithms, and Applications. CRC Press, Boca Raton, FL, USA.

Thirumalaiah K, M Deo 2000. Hydrological Forecasting Using Neural Networks. J Hydrol Eng, 5:180–189.

Thomas BF, A Behrangi, JS Famiglietti 2016. Precipitation Intensity Effects on Groundwater Recharge in the Southwestern United States. Water, 8(3):90.

Thorsteinsson T, T Johannesson, T Snorrason 2013. Glaciers and Ice Caps: Vulnerable Water Resources in a Warming Climate. Curr Opin Environ Sustain, 5(6):590–598.

Todd DK, LW Mays. 2005. Groundwater Hydrology (3rd ed., 656p). Wiley, Hoboken.

Todini E 1996. The ARNO Rainfall–Runoff Model. J Hydrol, 175:339–382.

Tong, STY, W Chen 2002. Modeling the Relationship between Land Use and Surface Water Quality. J Environ Manag, 66:377–393.

Treidel H, JL Martin-Bordes, JJ Gurdak. 2011. Climate Change Effects on Groundwater Resources: A Global Synthesis of Findings and Recommendations (1st ed.). CRC Press: London. https://doi.org/10.1201/b11611

Trisorio-Liuzzi G, A Hamdy 2008. Rainfed Agriculture Improvement: Water Management is the Key Challenge, 13th IWRA World Water Congress, Monpellier, France, 1–4 Sept. 2008.

Trnka M, R Brazdil, M Mozny et al. 2015. Soil Moisture Trends in the Czech Republic Between 1961 and 2012. Int J Climatol, 35:3733–3747.

Troell M, M Metian, M Beveridge et al. 2014. Comment on 'Water Footprint of Marine Protein Consumption-Aquaculture's Link to Agriculture. Environ Res Lett, 9:109001.

Tuyishimire A, Y Liu, J Yin et al. 2022. Drivers of the Increasing Water Footprint in Africa: The Food Consumption Perspective. Sci Total Environ, 809:152196.

Tuzet AJ 2011. Stomatal Conductance, Photosynthesis, and Transpiration, Modeling. In: Glinski J, Horabik J, Lipiec J, (eds) Encyclopedia of Agrophysics. Encyclopedia of Earth Sciences Series. Springer, Dordrecht. https://doi.org/10.1007/978-90-481-3585-1_213

Tyagi S, B Sharma, P Singh et al. 2013. Water Quality Assessment in Terms of Water Quality Index. Am J Water Resour, 1:34–38.

Uddin MG, S Nash, AI Olbert 2021. A review of Water Quality Index Models and their Use for Assessing Surface Water Quality. Ecol Indic, 122:107218.

Uhlenbrook S, W Yu, P Schmitter et al. 2022. Optimising the Water We Eat—Rethinking Policy to Enhance Productive and Sustainable Use of Water in Agri-Food Systems Across Scales. Lancet Planet Health, 6(1):e59–e65.

Ullah S, U Ali, M Rashid et al. 2024. Evaluating Land Use and Climate Change Impacts on Ravi River Flows Using GIS and Hydrological Modeling Approach. Scientific Reports, 14:22080.

United Nations Office for Disaster Risk Reduction 2021. GAR Special Report on Drought 2021. Geneva.

United Nations. 2019. World Population Prospects: The 2019 Revision. Department of Economic and Social Affairs, Population Division, United Nations. Online Edition. Rev. 1. Available at: https://population.un.org/wpp/Download/Standard/Population/.

Valencia JB, VV Guryanov, J Mesa-Diez et al. 2024. Predictive Assessment of Climate Change Impact on Water Yield in the Meta River Basin, Colombia: An InVEST Model Application. Hydrology, 11(2):25.

Valencia Cotera R, L Guillaumot, RK Sahu et al. 2023. An Assessment of Water Management Measures for Climate Change Adaptation of Agriculture in Seewinkel. Sci Total Environ, 885:163906.

Vallino E, L Ridolfi, F Laio 2020. Measuring Economic Water Scarcity in Agriculture: A Cross-Country Empirical Investigation. Environ Sci Policy, 114:73–85.

van der Ent RJ, L Wang, - Erlandsson, PW Keys et al. 2014. Contrasting Roles of Interception and Transpiration in the Hydrological Cycle—Part 2: Moisture Recycling. Earth Syst Dyn Discuss, 5:281–326.

van Dijk M, T Morley, ML Rau et al. 2021. A Meta-Analysis of Projected Global Food Demand and Population at Risk of Hunger for the Period 2010–2050. Nat Food, 2: 494–501. https://doi.org/10.1038/s43016-021-00322-9.

van Vliet M, GC Belleza, D Graham et al. 2024. Water scarcity Under Droughts and Heatwaves: Understanding the Complex interplay of Water Quality and Sectoral Water use, EGU General Assembly 2024, Vienna, Austria, 14–19 Apr 2024, EGU24-10512, https://doi.org/10.5194/egusphere-egu24-10512.

van Vliet MTH, J Thorslund, M Strokal et al. 2023. Global River Water Quality Under Climate Change and Hydroclimatic Extremes. Nat Rev Earth Environ, 4:687–702.

Vasilakou C, DE Tsesmelis, K Kalogeropoulos et al. 2025. Assessing Drought Severity in Greece Using Geospatial Data and Environmental Indices. Geomatics, 5(1):10.

Vásquez C, L Castillo, J Díaz et al. 2022. Analysis of Meteorological Droughts in the Sonora River Basin, Mexico. Atmósfera, 35:467–482.

Veettil AV, A Mishra 2020. Water Security Assessment for the Contiguous United States Using Water Footprint Concepts. Geophys Res Lett, 47:e2020GL087061.

Veettil AV, AK Mishra, TR Green 2022. Explaining Water Security Indicators Using Hydrologic and Agricultural Systems Models. J Hydrol, 607:127463.

Venkata Rao G, NR Nagireddy, VR Keesara et al. 2024. Real-Time Flood Forecasting Using an Integrated Hydrologic and Hydraulic Model for the Vamsadhara and Nagavali Basins, Eastern India. Nat Hazards, 120:6011–6039.

Verdegem M, AJT Dalsgaard, AH Buschmann et al. 2023. The Contribution of Aquaculture Systems to Global Aquaculture Production. J World Aquac Soc, 54:206–250.

Verhoeven JT, TL Setter 2010. Agricultural Use of Wetlands: Opportunities and Limitations. Ann Bot, 105(1):155–163.

Vesala T, S Sevanto, T Grönholm et al. 2017. Effect of Leaf Water Potential on Internal Humidity and Co2 Dissolution: Reverse Transpiration and Improved Water Use Efficiency Under Negative Pressure. Front Plant Sci, 8:54. https://doi.org/10.3389/fpls.2017. 00054.

Vishwakarma A, MK Choudhary, MS Chauhan 2024. Impact of Projected Climate Variability on Agricultural Water Demand of Key Crops in Drought-Prone Central India Region. J Water Clim Change, 15(9):4328–4342.

von Keyserlingk MAG, CJC Phillips, BL Nielsen 2016. Water and the Welfare of Farm Animals. In: Phillips, C. (eds) Nutrition and the Welfare of Farm Animals. Animal Welfare, vol 16. Springer, Cham. https://doi.org/10.1007/978-3-319-27356-3_9.

Vos J, AJ Haverkort. 2007. Water Availability and Potato Crop Performance. Potato Biol Biotechnol: Adv Perspect, 333–351. https://doi.org/10.1016/B978-044451018-1/50058-0

Wada Y, MF Bierkens 2014. Sustainability of Global Water Use: Past Reconstruction And Future Projections. Environ Res Lett, 9(10):104003.

Wada Y, LPH van Beek, CM van Kempen et al. 2010. Global Depletion of Groundwater Resources. Geophys Res Lett, 37:L20402.

Wada Y, LPH van Beek, CM van Kempen et al. 2010. Global Depletion of Groundwater Resources. Geophys Res Lett, 37(20):L20402. https://doi.org/10.1029/2010GL044571.

Wada Y, D Wisser, S Eisner et al. 2013. Multimodel Projections and Uncertainties of Irrigation Water Demand Under Climate Change. Geophys Res Lett, 40:4626–4632.

Wang B, LinHo 2002. Rainy Season of the Asian-Pacific Summer Monsoon. J Clim, 15:386–398.

Wang S, Q Zhao, T Pu 2021. Assessment of Water Stress Level About Global Glacier-Covered Arid Areas: A Case Study in the Shule River Basin, Northwestern China. J Hydrol Reg Stud, 37:100895.

Wang W, E Straffelini, P Tarolli 2024. 44% of Steep Slope Cropland in Europe Vulnerable to Drought. Geogr Sustain, 5:89–95.

Wang Y, B Geerts, C Liu et al. 2025. A Convection-Permitting Regional Climate Simulation of Changes in Precipitation and Snowpack in a Warmer Climate over the Interior Western United States. Climate, 13:46.

Wanyama J, E Bwambale, S Kiraga et al. 2024. A Systematic Review of Fourth Industrial Revolution Technologies in Smart Irrigation: Constraints, Opportunities, and Future Prospects for Sub-Saharan Africa. Smart Agric Technol, 7:100412.

Ware HH, TD Mengistu, BA Yifru et al. 2023. Assessment of Spatiotemporal Groundwater Recharge Distribution Using SWAT-MODFLOW Model and Transient Water Table Fluctuation Method. Water, 15(11):2112.

Warziniack TM, TC Arabi, P Brown et al. 2022. Projections of Freshwater Use in the United States Under Climate Change. Earth's Future, 10(2):1–20.

Wasko C, C Stephens, TJ Peterson et al. 2024. Understanding the Implications of Climate Change for Australia's Surface Water Resources: Challenges and Future Directions. J Hydrol, 645:132221.

Weyant C, ML Brandeau, M Burke et al. 2018. Anticipated Burden and Mitigation of Carbon-Dioxide-Induced Nutritional Deficiencies and Related Diseases: A Simulation Modeling Study. PLoS Med, 15:e1002586.

White DJ, K Feng, L Sun et al. 2015. A Hydro-Economic Mrio Analysis of the Haihe River Basin's Water Footprint and Water Stress. Ecol Modell, 318:157–167.

Whitehead PG, L Jin, G Bussi et al. 2019. Water Quality Modelling of the Mekong River Basin: Climate Change and Socioeconomics Drive Flow and Nutrient Flux Changes to the Mekong Delta. Sci Total Environ, 673:218–229.

WHO. 2021. Drought Overview. WHO Website. https://www.who.int/health-topics/drought#.

Wichelns D 2015. Water Productivity and Food Security: Considering more Carefully the Farm-Level Perspective. Food Sec, 7:247–260.

Wilmsen B, M Webber 2017. Mega Dams and Resistance: The Case of the Three Gorges Dam, China. In: Grugel, J, J Nem Singh, L Fontana, A Uhlin (eds), Demanding Justice in The Global South. Development, Justice and Citizenship. Palgrave Macmillan, Cham. https://doi.org/10.1007/978-3-319-38821-2_4.

WMO 2020. State of the Global Climate 2020. World Meteorological Organization. https://library.wmo.int/index.php?lvl=notice_display&id=21880#.YHg0ABMzZR0

Wolkeba FT, M Kumar, MM Mekonnen 2023. Examining the Water Scarcity Vulnerability in US River Basins Due to Changing Climate. Res Lett, 50(24):e2023GL106004.

Wu B, F Tian, M Zhang et al. 2022. Quantifying Global Agricultural Water Appropriation with Data Derived From Earth Observations. J Clean Prod, 358:131891.

Wu XH, W Wang, CM Yin et al. 2017. Water Consumption, Grain Yield, and Water Productivity in Response to Field Water Management in Double Rice Systems in China. PLoS One, 12(12):e0189280.

Wudil AH, M Usman, J Rosak-Szyrocka et al. 2022. Reversing Years for Global Food Security: A Review of the Food Security Situation in Sub-Saharan Africa (SSA). Int J Environ Res Public Health, 19(22):14836.

Xanke J, T Liesch 2022. Quantification and Possible Causes of Declining Groundwater Resources in the Euro-Mediterranean Region From 2003 to 2020. Hydrogeol J, 30: 379–400.

Xinchun C, W Mengyang, G Xiangping et al. 2017. Assessing Water Scarcity in Agricultural Production System Based on the Generalized Water Resources and Water Footprint Framework. Sci Total Environ, 609:587–597.

Xu C, E Widén, S Halldin 2005. Modelling Hydrological Consequences of Climate Change— Progress and Challenges. Adv Atmos Sci, 22:789–797.

Yadav M, BB Vashisht, SK Jalota et al. 2024. Improving Water Efficiencies in Rural Agriculture for Sustainability of Water Resources: A Review. Water Resour Manage, 38:3505–3526.

Yan D, M Yao, F Ludwig et al. 2018. Exploring Future Water Shortage for Large River Basins Under Different Water Allocation Strategies. Water Resour Manag, 32(9):3071–3086.

Yáñez-Arancibia A, JW Day 2017. Water Scarcity and Sustainability in the Arid Area of North America: Insights Gained from a Cross Border Perspective. Reg Cohes, 1:6–18.

Yang D, Y Yang, J Xia 2021. Hydrological Cycle and Water Resources in a Changing World: A Review. Geogr Sustain, 2(2):115–122.

Yildirim Ü, C Güler, B Önol et al. 2021. Modelling of the Discharge Response to Climate Change under RCP8.5 Scenario in the Alata River Basin (Mersin, SE Turkey). Water, 13(4):483.

Yimer EA, FE Riakhi, RT Bailey et al. 2023. The Impact of Extensive Agricultural Water Drainage on The Hydrology of The Kleine Nete watershed, Belgium. Sci Total Environ, 885:163903.

Yin W, X Yang, W Liu 2025. Sustainable Management and Regulation of Agricultural Water Resources in the Context of Global Climate Change. Sustainability, 17(6):2760.

Young S, E Frongillo, Z Jamaluddine, et al. 2021. Perspective: The Importance of Water Security for Ensuring Food Security, Good Nutrition, and Well-Being. Adv Nutr. https://doi.org/ 10.1093/advances/nmab003

Young SL, EA Frongillo, Z Jamaluddine et al. 2021. Perspective: The Importance of Water Security for Ensuring Food Security, Good Nutrition, and Well-Being. Adv Nutri, 12(4):1058–73.

Young SS 2023. Global and Regional Snow Cover Decline: 2000–2022. Climate, 11(8): 162.

Yuan W, Q Liu, S Song et al. 2023. A Climate-Water Quality Assessment Framework for Quantifying the Contributions of Climate Change and Human Activities to Water Quality Variations. J Environ Manag, 333.

Zain Eldin AM, ME Attia, TK Zin El-Abedin et al. 2024. Effect of Magnetized Saline Irrigation Water on Soil Mechanical Properties, Emitters Efficiency and Yield of Eggplant in Saline Soils. Misr J Agric Eng, 41:39–62.

Zapata-Sierra AJ, L Zapata-Castillo, F Manzano-Agugliaro 2022. Water Resources Availability in Southern Europe at the Basin Scale in Response to Climate Change Scenarios. Environ Sci Eur, 34:75.

Zarrineh N, KC Abbaspour, A Holzkämper 2020. Integrated Assessment of Climate Change Impacts on Multiple Ecosystem Services in Western Switzerland. Sci Total Environ, 708:135212.

Zeggaf Tahiri A, G Carmi, M Ünlü 2021. Promising Water Management Strategies for Arid and Semiarid Environments. In: Landscape Architecture - Processes and Practices Towards Sustainable Development. IntechOpen: London. Available from: http://dx.doi.org/10.5772/intechopen.87103

Zetland D 2021. The Role of Prices in Managing Water Scarcity. Water Secur, 12:100081.

Zewde NT, MA Denboba, SA Tadesse et al. 2024. Predicting Runoff and Sediment Yields Using Soil and Water Assessment Tool (SWAT) Model in the Jemma Subbasin of Upper Blue Nile, Central Ethiopia. Environ Chall, 14.100806.

Zhang L, GR Walker, WR Dawes 2002. Water Balance Modelling: Concepts And Applications. ACIAR Monograph Series, 84:3147.

Zhang Q, SL Wang, YG Sun et al. 2022. Conservation Tillage Improves Soil Water Storage, Spring Maize (*Zea mays* L.) Yield and WUE in Two Types of Seasonal Rainfall Distributions. Soil Tillage Res, 215:105237.

Zhang S, YF Sang, T Qiu et al. 2024. Water Resource Availability and Use in Mainland Southeast Asia. In: Chen, D, J Liu, Q Tang (eds), Water Resources in the Lancang-Mekong River Basin: Impact of Climate Change and Human Interventions. Springer, Singapore. https://doi.org/10.1007/978-981-97-0759-1_5.

Zhao C, B Liu, S Piao et al. 2017. Temperature Increase Reduces Global Yields of Major Crops in Four Independent Estimates. Proc Natl Acad Sci USA, 114(35):9326–9331.

Zhao Q, L Wu, F Huo et al. 2025. Assessing the Impacts of Shifting Planting Dates on Crop Yields and Irrigation Demand Under Warming Scenarios in Alberta, Canada. Agric Water Manag, 309:109304.

Zhao RJ, XR Liu 1995. The Xinanjiang Model. In: Singh, VP (ed), Computer Models of Watershed Hydrology (pp. 215–232). Water Resource Publications, Highlands Ranch.

Zhongwei H, M Hejazi, Q Tang et al. 2019. Global Agricultural Green and Blue Water Consumption Under Future Climate and Land Use Changes. J Hydrol, 574:242–256.

Zhu Y, Z Cheng, K Feng et al. 2022. Influencing Factors for Transpiration Rate: A Numerical Simulation of an Individual Leaf System. Therm Sci Eng Prog, 27:101110.

Zinkernagel J, JF Maestre-Valero, SY Seresti et al. 2020. New Technologies and Practical Approaches to Improve Irrigation Management of Open Field Vegetable Crops. Agric Water Manag, 242:106404.

Zisopoulou K, D Panagoulia 2021. An in-Depth Analysis of Physical Blue and Green Water Scarcity in Agriculture in Terms of Causes and Events and Perceived Amenability to Economic Interpretation. Water, 13(12):1693.

Zolghadr-Asli B, O Bozorg-Haddad, M Enayati, et al. 2021. A Review of 20-Year Applications of Multi-Attribute Decision-Making in Environmental and Water Resources Planning and Management. Environ. Dev. Sustain, 23(10), 14379–14404. https://doi.org/10.1007/s10668-021-01278-3.

Zribi W, R Aragüés, E Medina et al. 2015. Efficiency of Inorganic and Organic Mulching Materials for Soil Evaporation Control. Soil Tillage Res, 148:40–45.

Zwart SJ, WGM Bastiaanssen 2004. Review of Measured Crop Water Productivity Values for Irrigated Wheat, Rice, Cotton, and Maize. Agric Water Manag, 69(2):115–133.

Zyngier RL, CL Archibald, BA Bryan et al. 2024. Knowledge Co-Production for Identifying Indicators and Prioritising Solutions for Food and Land System Sustainability in Australia. Sustain Sci, 19:1897–1919.

Index